T0179831

The Visualization of Spatial Social Structure

WILEY SERIES IN COMPUTATIONAL AND QUANTITATIVE SOCIAL SCIENCE

Embracing a spectrum from theoretical foundations to real world applications, the Wiley Series in Computational and Quantitative Social Science (CQSS) publishes titles ranging from high level student texts, explanation and dissemination of technology and good practice, through to interesting and important research that is immediately relevant to social/scientific development or practice.

The Visualization of Spatial Social Structure

Daniel Dorling

Department of Geography, University of Sheffield, UK

A John Wiley & Sons, Ltd., Publication

Registered office
John Wiley & Sons Ltd, The Atrium, Southern Gate, Chichester, West Sussex, PO19 8SQ, United Kingdom

For details of our global editorial offices, for customer services and for information about how to apply for permission to reuse the copyright material in this book please see our website at www.wiley.com.

Library of Congress Cataloging-in-Publication Data

Dorling, Daniel.
  The visualization of spatial social structure / Daniel Dorling.
     p. cm.
  Includes bibliographical references and index.
  ISBN 978-1-119-96293-9 (cloth)
  1. Human geography–Great Britain. 2. Cartography–Methodology. 3. Cartography–Philosophy. I. Title.
  GF551.D674 2005
  304.2072'8–dc23

                                                                            2012009924

A catalogue record for this book is available from the British Library.

ISBN: 978-1-119-96293-9

Typeset in 10/12pt Times by Laserwords Private Limited, Chennai, India
Printed and bound in Singapore by Markono Print Media Pte Ltd

To Benjamin Dorling (1971–1989)

Contents

Note. The original thesis from which this book was derived had a further six appendices and a larger bibliography. Some can be found at www. dannydorling.org.

Appendix A: Circular Cartogram Algorithm

Appendix B: Parliamentary Constituencies 1955–1987 Continuity

Appendix C: Parliamentary Constituencies 1955–1987 Results

Appendix D: Average Housing Price by Constituency 1983–1989

Appendix E: Scottish Ward to Postcode Sector Look-up Table

Appendix F: Local Government Wards, 1981 and 1987

List of figures

List of text boxes

Preface

This book tells a story about seeing things differently. The story is a way of introducing the reader to new ways of thinking about how to look at social statistics, particularly those about people in places.

The visualization of spatial social structure means, literally, trying to make visible the geographical patterns to the way our lives have come to be socially organised, seeing the geography in society. To a statistical readership visualization implies using data. More widely defined it implies freeing our imaginations.

The story of this book centres on a particular place and time, 1980s Britain, and a particular set of records, routine social statistics. A great deal of information about the 1980s social geography of Britain is contained within databases such as the population censuses, surveys and administrative data. During the 1980s computer graphics developed and, to comprehend the information they held, a few social scientists thought it needed to be effectively visualized with computer graphics (Figure P.1).

In the United States a small but significant number of geographers in the 1960s[1] argued that conventional maps contained a massive and unwanted distortion, but a growing number in the social sciences back in the 1970s then thought that anything numerical was in some way suspicious and could de-humanise inquiry. This work builds on listening to the latter, but also on developing the techniques of the former group, which have been largely ignored in the 1990s and the 2000s.

Mapping, by the late 1980s, had been rejected by many social analysts as an unsuitable means of showing spatial social structure. The usual alternative was, and remains, to write in the abstract on social structure and rarely to employ graphics or maps or to rely on numbers. However, that wastes a huge amount of information and the skills of many more numerically minded people who might also be interested in uncovering the social organisation of the world they live in.

A human cartography is proposed here that reveals, through amalgamating and subdividing the events of people's lives, the shape of society (Figure P.2). The aim here is to see the whole, in as much detail as possible, at a glance. While the case study is 1980s Britain, the geography of Thatcherism, the applicability of these techniques is hopefully far wider. The areas studied could be far smaller

---

[1] The work of one was recently republished (Bunge, 2011).

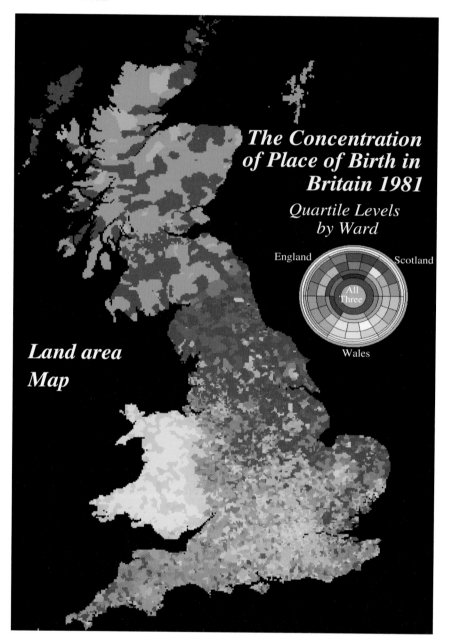

Figure P.1   In the 1980s ward data might be put on a grid of small squares as shown here. Each ward is coloured by the proportions of people born in each of the three main countries of Britain, but who are now living in that ward. The mixing of colours suggests the outcome of lifetime migration patterns. However, the map is misleading, overemphasising mixing in remoter rural areas.

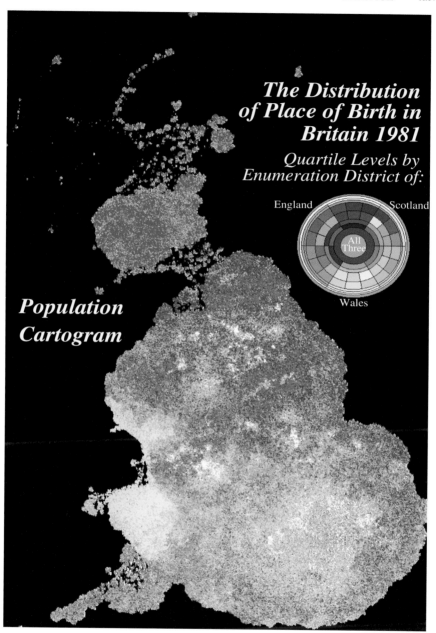

Figure P.2   This population cartogram shows the mix of people by birthplace in 129 211 small areas of similar population size. All are visible. On the previous map even most wards are invisible. Here it is clear that neighbourhoods were not mixed in much of urban Scotland and Wales. Areas coloured white – where more people are born overseas – can now be seen.

than an island like Britain, or larger. Revealed here is the society inherited by Margaret Thatcher's government in 1979 and how that society had been changed by 1990, the year of her forced resignation.

These same techniques could be applied to visualize a state like California from when it was dreaming in 1965 to when it was potentially bankrupt in 2012. The more human focused forms of cartography proposed here include new ways of looking at the geography and social statistics of places as large as India, as remote as Anchorage or even as tiny as number 29 Acacia Avenue.

The illustrations included here are what is core to this work. They include pictures of the distribution of age, sex, birthplace and occupation across Britain in 1981, changes in these from 1971, unemployment and house price dynamics throughout the 1980s boom and 1989 bust, general election results from 1955 to 1987 (followed by all local election voting from 1987 to 1990), visual summaries of migration flows from one part of the country to another and drawings of thousands of daily commuting streams (Figure P.3).

The creation of simple computer generated cartograms is explained, where each spatial unit (up to one hundred thousand to a page) is drawn with its area proportional to the number of people who live there. Colour and complex symbols are used to study several factors simultaneously upon these cartogram bases, to let the analyst compare different datasets at the same time, for what they show about the same places.

Novel visually effective means of showing millions of flows and other changes over time are also developed (Figure P.4). Further advances are imagined and travel time surfaces are described, through which tunnels are cut and over which other information can be draped. A case study of the distribution of childhood leukaemia in space and time is undertaken, showing a pattern of no pattern[2] (Figure P.5). The detailed results of the ten general elections up to Margaret Thatcher's last victory, of 1987, are compared. Revealing images of the beginnings of how Britain came to be set on the path to growing polarisation is a theme that runs throughout.

Essentially, however, this is a book about graphical techniques, not about social history, epidemiological analysis or political study. Twenty years ago almost no visualization software existed. To draw a map required writing a computer program. This meant you could draw a map in many different ways. Today software has become sufficiently versatile that, without needing to program, it might *again* be possible to produce the kind of images you might want to produce, rather than those you might get from the default options.

This book is about a spatial way of thinking of the structure of society – of social structure – and how you might draw what it is you are thinking of, if you think of it in a particular way. Although it uses examples from the past, the focus

---

[2] This is a lack of clustering later confirmed in numerous studies with access to many more years of data. In May 2011 '... there is no evidence to support the view that there is an increased risk of childhood leukaemia and other cancers in the vicinity of NPPs (Nuclear Power Plants) in Great Britain' (Elliott, 2011, p. 102).

here is on technique, not subject. However, the particular past is of great interest to some (including this author who cannot resist making asides as a result). Prior to 2011, the 1980s were the last time Britain faced mass unemployment, rapidly growing social divisions or widespread rioting.

This text describes the rationale for, and development of, a new way of visualizing information in geographical research (Figure P.6). Through the pictures the methods are illustrated and mistakes, techniques and discoveries shown. From the footnotes, which are largely quotations from a disparate literature, the origins of many of the ideas can be found. Time and again it was the suggestions of others to move in these directions.

Through technical asides some of the practical realities of the work are described. Through the illustrations and their captions, a picture of what had been happening to Britain in those recession years unfolds. Many of the pictures could justify an extended discussion, but the commentary is kept brief. Little detail is included about the computer software written and used here because much of that is dependent on the novel (but inexpensive) Acorn hardware configuration and progress is so rapid that such knowledge is of only transient value.[3]

The images in this book reveal how in aggregate people get to work and the structure of the towns and cities in which they live is examined. Migration (moving home) is studied here in several ways. The changing patterns of migration from birthplaces are shown and the streams of movement that cut across the country are drawn in unprecedented, and as yet not superseded, detail.

House price change is visualized across several years and thousands of places. This detail reveals that the origins of the 1989 crash lay years earlier in the heart of the London housing market. Other new techniques are developed to show the structure of local housing markets. Through different methods again, the changes in this country's industrial structure are seen as they have affected people in actual communities.

Figure P.7 shows a simplified version of a more complex chart of industrial change in which just eight industrial groups are shown, but also how those groups of employment altered for men and women, and for full- and part-time workers separately in 1981, 1984, 1987 and between each consequent pair of survey years.

This single small collection of five glyphs suggests that male full-time manufacturing job losses in the early 1980s were not quite replaced by female full- and part-time work, mostly in service industries, mostly later on in the 1980s. A lot can be shown in just a few simple images.

The spatial and social manoeuvring of political allegiances is viewed from several angles over the same period and the relationships discussed. Finally, a smaller scale of analysis is considered, looking at what many images can tell us about the distribution of a disease, viewed from many different directions in space and time.

---

[3] An Acorn Archimedes computer was used, produced by the company known as the 'British Apple', which existed for twenty years from 1978 to 1998. See the endnote to this book for more details.

Yearly Migration Flows Between English and Welsh Wards 1980/1981.

32% of all migrants included 1,352,520 people.

Flows of more than 0.2% of the geometric mean for the resident populations of the areas of origin and destination are drawn as thin lines, shown on an **land area map.**

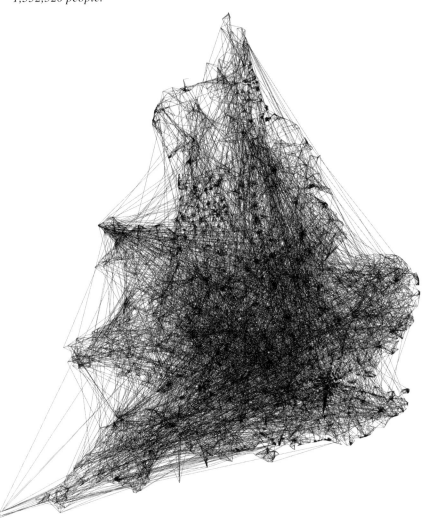

Figure P.3   Each line represents a minimal number of moves made between wards in England and Wales in one year on a conventional land area map. The interward migration patterns show a complete tangle of lines. The Isle of Wight can be made out, as can the outlines of some towns and cities, but in general areas towards the centre of England simply become mostly criss-crossed.

Yearly Migration Flows Between English and Welsh Wards 1980/1981.

32% of all migrants included 1,352,520 people.

Flows of more than 0.2% of the geometric mean for the resident populations of the areas of origin and destination are drawn as thin lines, shown on a **_population cartogram._**

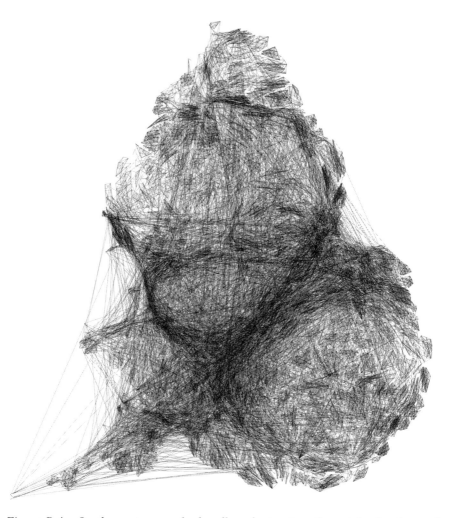

Figure P.4   On the cartogram the bundles of migratory flows take the shape of London boroughs and other areas from which council house tenants have found it difficult to move in the past. The more prosperous areas of the country are black-ened by the density of flows in and through them. The shape of the conurbations is clear, as people who can avoid living there migrate around them.

Selected Frames from an Animation of the Spacetime Distribution of Childhood Leukaemia

Across the North of England from 1967 to 1987, upon an equal land area projection

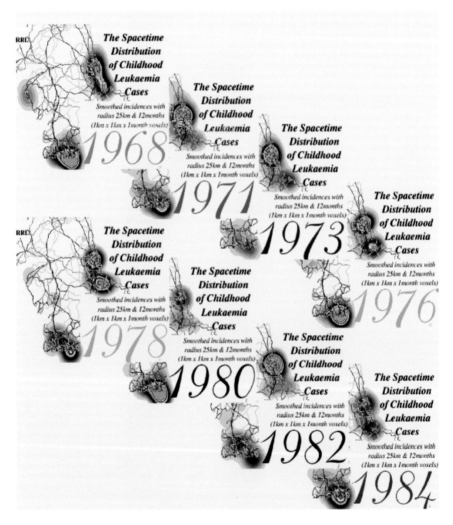

Figure P.5   These eight frames are taken from an animation of the changing concentration of cases of childhood leukaemia (where time is the moving third dimension). They define a volume within which rates are estimated and smoothed. Although it appears there is clustering, the methods used tend to find the areas of highest population density when the base is a land area map. Note: parts of original image were produced as a bit-map of pixels of colour not as a vector graphics file of lines, curves and areas.

Selected Frames from an Animation of the Spacetime Distribution of Childhood Leukaemia

Across the North of England from 1967 to 1987, using population space.

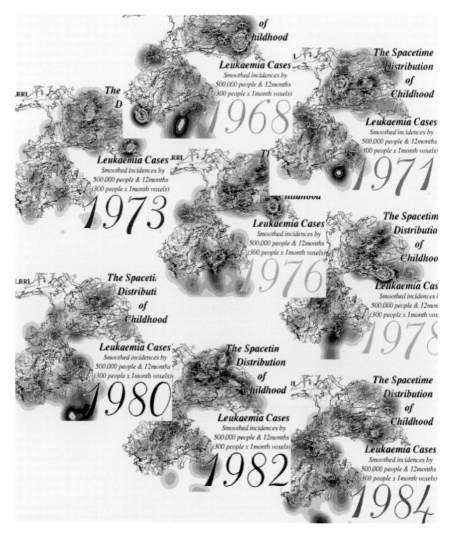

Figure P.6   These eight frames are taken from an animation of the changing concentration of cases of childhood leukaemia over equal population sized areas defining life volumes within which rates are estimated and smoothed. Although it still appears there is clustering the methods used would draw apparent but forever moving concentrations, even from randomly generated data.

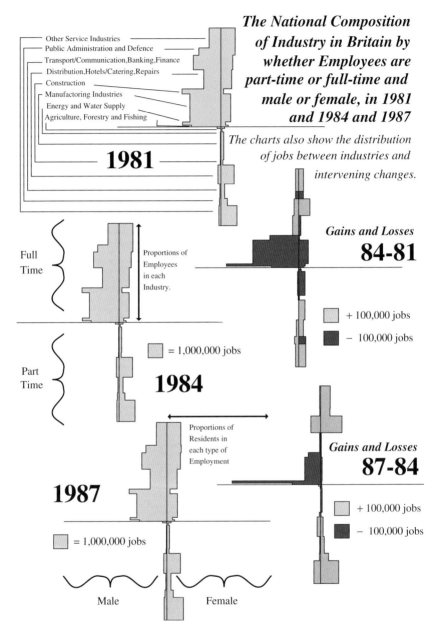

Figure P.7 In the 1980s it was common to aggregate banking and finance with the communication and transport industries as these were all seen as facilitating other work, not as profit-making centres in their own right. Note how so many more jobs were lost in the first period of the 1980s, 1981–1984, as compared to the second period, 1984–1987.

The National Composition of Housing Price in Britain by Price, Attributes and Sales, 1983-1989 anually.

Each branch of the housing composition tree represents homes with a particular set of attributes. The width of each branch is in proportion to the number of sales, its length gives the mean price, thus area shows total sales. Dark trees are the inflation between years (magnified by 25). The overall size of each tree illustrates the total size of the housing market (given in millions of pounds).

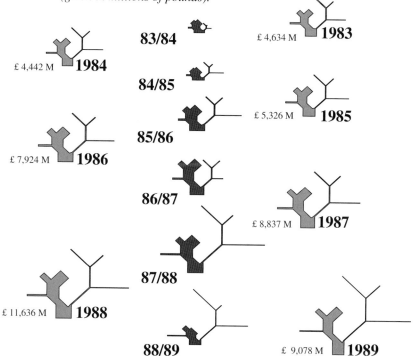

Figure P.8   In these glyphs, the housing market is divided into a number of branches, the branches join to show bigger sectors and the trunk represents the whole housing market. The shape of the resultant trees shows the housing market structure. Here the national shapes are shown. Inflation causes the length of a branch to increase and fewer sales in a sector cause the branch to become thinner.

These social and political subjects are not each arranged in their own, individual chapters, but run through the book, as it is a book about new possible methods of visualization rather than the visualization of subjects. The rationale for using images to study people, places and spaces is discussed as the new images are introduced.

The central part of the book looks at what appears to be a honeycomb structure formed by a particular method of viewing the spatial patterns of society at single points in time and how that image alters through transforming the envisioned mosaic. The cobweb of flows that is responsible for most of the changes and stability is then drawn.

The last part of this book attempts to show more complex aspects on the surface of social landscapes. Sculptured symbols allow us to see the relationships between the wood and the trees of social structure (Figure P.8). Finally, a three-dimensional volume visualization of geographical and historical social structure, of spacetime, is attempted. The book concludes by describing how all these methods and insights can, when brought together, create a new statistical view of human geography and recent history.

Visualization in the social sciences demands that we consider what is happening in many places at the same time. We do not need to study aspects of the world out of context. Here, an attempt is made to cover much ground and show numerous relationships. To do this it is necessary to be brief in detail and to be broad in scope, so the pictures often have to speak for themselves. Only once you have seen what it is you want to talk about can you then better ask questions and make interpretations.[4]

---

[4] 'The analytic power to order data has potential equally for control or liberation. It is all a matter of questions asked and interpretations made' (Taylor, 1991, p. 30).

Introduction: Human cartography

Images are only images. But if they are numerous, repeated, identical, they cannot all be wrong. They show us that in a varied universe, forms and performances can be similar: there are towns, routes, states, patterns ... which in spite of everything resemble each other.

(Braudel, 1979, p. 133)

This book presents the thesis that light can be cast on the study of society through the visualization of social structure. The antecedents of the work presented here lie most firmly in human geography and cartography while being influenced by writings in other disciplines. There are contributions from studies in computer and statistical graphics, graphic design and art, mathematical abstraction (Figure I.1) and political science.[1]

Particular views on the study of history, geography and sociology guide much of the writing. Above all, this book is concerned with designing and advocating new ways of seeing the social world we live in. Before doing that, it is necessary to explain why still widely accepted graphical techniques are being discarded by the visual methodology proposed here. Most important of all, in order to show the spatial structure of society the conventional use of maps of physical geography has to be rejected.

Maps were designed to explore new territories and fight over old ones. They show where oceans lie and rivers run. Their projections are calculated to aid navigation by compass or depict the quantity of land under crops (Figure I.2). They are a flat representation of part of the surface of the globe; they show things that often cannot be seen. How then can we see social structure, in the same manner as the map opens up land to the eye?[2] How can we begin to see

---

[1] See Muehrcke (1978, 1981), Szegö (1984, 1987), Anderson (1988) and Cuff (1989).

[2] The advantage of maps is simple – they provide context: 'Maps frown upon the isolation of single items. They preserve the continuity of the real world. They show things in their surroundings and therefore call for more active discernment on the part of the user, who is offered more than he came for; but the user is also being taught how to look at things intelligently. One aspect of looking at things intelligently is to look at them in context' (Arnheim, 1976, p. 5).

Figure I.1   Convergence of $z := z^2 + c$ on the complex plane. The Mandelbrot set exhibits infinite variations on the same visual theme. Every picture is familiar, but all are subtly different. As you magnify the image, the detail and variation is as great at every level. In human geography the same degree of diversity can often be recognised on a local scale, as can be seen nationally.

*Land Use
Close Up.*

Urban
Agriculture
Open Land
Woodland
Moorland
Upland
Inland Water

*Figure I.2   The image is centred on the land-use pattern of the North East of
England. Tyneside, Wearside and Carlisle can be distinguished, as can the pattern
of farmland and other landscapes. Just over thirty years ago this was some of the
most high resolution data available (from researchers at Merlewood). Population
was also mapped by grid-square in the late 1970s.*

the patterns of society, which we know are there since we help form them, but patterns that we may have never literally viewed?

Maps were not designed to show the spatial distributions of people, although the single spatial distribution of people upon the surface of the globe at one instance in time can be shown on them. They cannot illustrate the simplest human geography of population. People are but points on the conventional map, clustered into collections of points called homes, into groups of points known as villages or cities. Communities of people are not like fields of crops. The paths through space that they follow are not long wide rivers of water and yet, to see anything on maps of people, they must be shown as such.

Conventional maps cannot show how many people live in small areas; instead they show how little land supports so many people. They cannot show who the people are, what they do, where they go. They show no temporal distribution, they do not need to – how quickly do rivers move or mountains shrink on a human timescale? They will not be an appropriate base to show the distributions of people changing – international migration, moving house or just people going to work (Figure I.3).

The aim here is to make sense of the reality of thousands of people simultaneously threading their way through life. What are they doing and why are they doing it? How can we see into every home, know what everyone does? We can't, but we can guess and we have some clues. We can guess from what, introspectively, we know from being part of society. We amass clues when people are counted.

There has been an obsession for counting people since recording became possible. Every ten years, in many countries, hundreds of thousands of people count people (in the census). Increasingly our actions are being recorded; we are now each noted several times a day, from the heat we register on satellite images to almost every transaction and phone call we make or unit of electricity we consume.

The conventional statistical treatment of numerical information about people averages them, agglomerates them and destroys the detail that is of interest,[3] taking a million numbers and returning just half a dozen. These techniques were conceived when little better could be done. Now it is possible to show you a million bits of information at a glance that would be challenging to describe in a thousand words (Figure I.4).

Pictures can make ideas plausible, the screen beautiful, millions of numbers meaningful. They have intrigued many, as maps and charts of rivers and mountains.[4] Here traditional maps are the inspiration, but not the foundation, for the

---

[3] 'The dismissal of geographical diversity as merely "noise" or "residuals" is a betrayal of what geography is' (Taylor, 1991, p. 24).

[4] An old definition is surprisingly apt: 'One of the definitions of the word "map" that appears in the Oxford Dictionary dates from a source of 1586, where it was used to describe "a circumstantial account of a state of things" ("circumstantial" is defined as "full of circumstances, details or minutiae"): not a bad objective 400 years later!' (Bickmore, 1975, p. 328).

Yearly Migration Flows between English and Welsh Wards from 1980 to 1981

on a population cartogram.

All flows of over 1 in 100 of the origin ward plus destination ward population.
(5.2% of the total number migrating between wards.)

The single flow to Scotland involved over 1% of the population of the ward of origin.

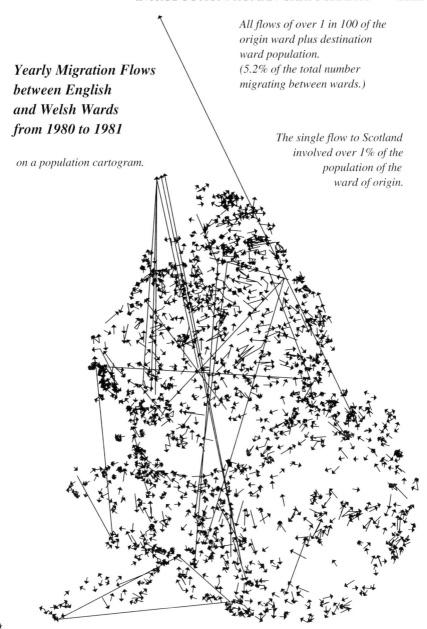

Figure I.3   It might be thought that reducing the amount of information shown on a migration map could clarify the picture, but here an image is drawn suggesting that is not the case. When just the largest flows are included these mostly tend to be between neighbouring wards that share a particularly wiggly boundary and/or where the wards have larger populations.

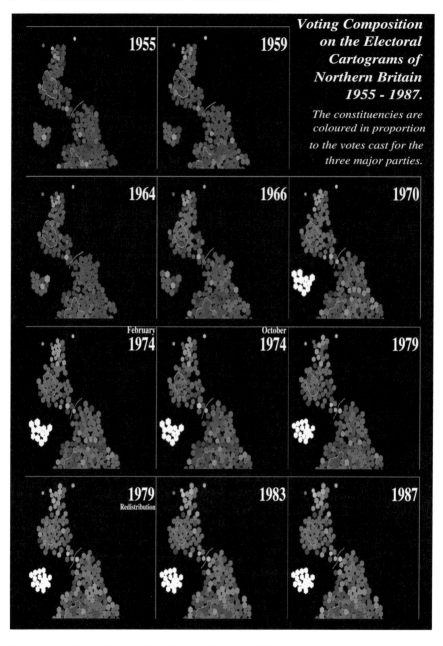

Figure I.4   Votes for the Northern half of the 705 parliamentary constituencies. These are the key frames to a video animation of the change in voting over ten elections. The basic shape of allegiances in the North can be seen to have remained much the same over the period. The most dramatic change is in Northern Ireland's exit from the main political arena.

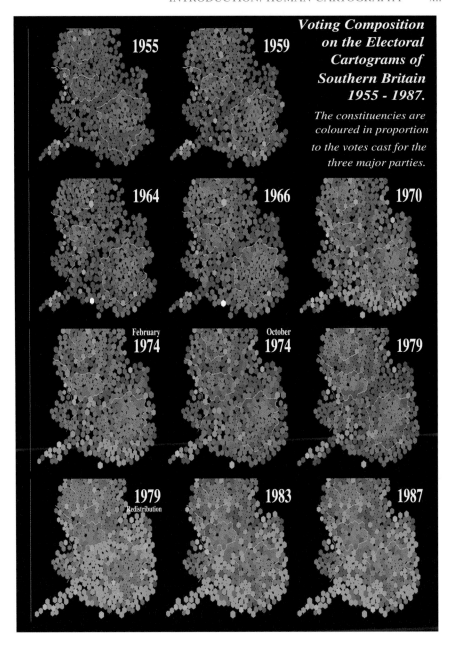

Figure I.5   In the South of the country we see the most dramatic changes in people's voting patterns. From a fairly mixed first image in the 1950s, a divided nation had developed by 1987. The darkening ring of greeny-blue (Liberal and Conservative) constituencies around the Capital is clear, as is the geography of its evolution. The boundaries of the major conurbations are shown by faint lines.

creation of new graphics to form pictures of people, with their rivers of roads down which they flow and mountains of cities up which they climb.

How millions of human beings cooperate, compete and coordinate can be very hard to grasp, let alone understand how they then interact with other objects, species and resources. Our minds are the most powerful tools we have to address these problems. The difficulty comes in trying to address these problems to our minds.

From 1975 to 1989 orthodox cartographic methodology was translated on to the computer screen.[5] The subject of cartography was subsumed by the study of geographic information systems, but the fundamental basis to cartography, physical geography, has remained.[6] Thematic maps drastically distort the reality they purport to contain, at worst reversing the patterns that exist.

People who study people, who are interested in societies, politics, history, economics and increasingly even those interested in human geography, do not often use maps or other graphics. A topographic map base allows, at most, the depiction of human land use. People have created maps based on human geography in the past, but only with the advent of sophisticated computer graphics, visualization, has it become possible to do this on an easily replicable basis.

Most important is addressing the problem of how time and space can be transformed to represent clearly the patterns within them, on paper or in animation. Transformation is inherent in representation (Figure I.5). That is a most difficult concept to accept. It inevitably affects the images produced and the emphasis the viewer places on different places and, more importantly, the relationship between places and times – the metric. The argument for transforming has been made repeatedly over many decades, but usually by mavericks who have been mostly ignored. For instance, in 1966 Bill Bunge wrote:

> *Consider the use of rivers on base maps. With the invention of bridges to cross them and railroads to compete with transportation on them, it could be argued that rivers have become unimportant enough to be eliminated from the map. They might be replaced by major railroad lines. ... Major cities are more important 'islands' for many purposes than the atolls of the Pacific. It is probably true that of all the degrees of latitude and longitude shown on the map, only the equator and the poles are on the mental map and, therefore, the other degrees might be dropped as superfluous. ... It is much easier to plot the continental outlines, rivers and mountain peaks than to obtain a census of population or an accurate map of arable land.*
>
> *(Bunge, 1966, pp. 45–46)*

---

[5] See Bickmore (1975), Taylor (1985), Goodchild (1988) and Muller (1989).

[6] '... digital cartography ... can successfully emulate its analogue parent. However, its true potential lies in less conventional methods of analysis and display and in the degree to which it can escape its traditional constraints' (Goodchild, 1988, p. 311).

William Bunge has recently been rediscovered, and is now presented as being one of the most pioneering social scientists of the 1960s.[7] Many of his ignored claims are reiterated in stronger terms here. They call for new images, most especially cartograms.

Put simply, on a cartogram people no longer exist on paper as points, but as areas, so they can now be legitimately drawn (when grouped) as fields (Figure I.6); their paths of aggregate movement appear now as rivers perhaps running through a landscape of accessibility covered with the vegetation of some aspect of social structure.[8]

In many countries the clues to social structure given by official administrative sources and in survey sources consist mostly of the absolute numbers, such as the age and sex of people across the country. Then, every ten years the combinations of their answers to a few questions at the census are provided – where they were born, what job they do or did, where they did it, where they moved. Here only British data are used as a case study to illustrate how much possible visualization there is, even of one small island.

There are many noncensus forms of information that can be drawn on (Figure I.7). One claim of visualization is the ability to handle large quantities of loosely related data coherently. Some other sources and surveys are called upon here, but all from a particular era. Although this work was first drafted between 1989 and 1991, it has been extensively redrafted given the hindsight of twenty years. Much of the text is new, but all the examples are taken from then or before to constrict the case study in time as well as space.

Alternative sources to the routine administrative data used in these pages include how people voted in the local and national elections of the decades being studied (Figure I.8), national surveys of workplaces that were conducted in several years in the 1980s in Britain, the health service records of internal migration, building societies' lists of house sales and information on the infrastructure of roads, railways and settlements. These were digitally available in the 1980s. What is shown here is the means of putting these numbers together, as a collection of images forming one picture of one place during a short, dozen years or so, period of time.

The simplest of spatial distributions to envisage are those captured at single instances of time, and so it is with these that this book begins. Later chapters show that it is possible to visualize the changes in the population over time in tens of thousands of streets within a single picture. Much of what this shows about Britain will be unexpected even though it is not new information that is being used here and even though there has been much time for other ways of study to uncover what is shown here.

---

[7] See: http://indiemaps.com/blog/2010/03/wild-bill-bunge/ by Zachary Forest Johnson, 16 March 2010.

[8] 'In physical geography, only that which has an effect on mankind is studied. Now that men are much less dependent on the countryside than on cities, why have geographers not followed mankind? Why have geographers left their minds back on the farm?' (Bunge, 1975, p. 177).

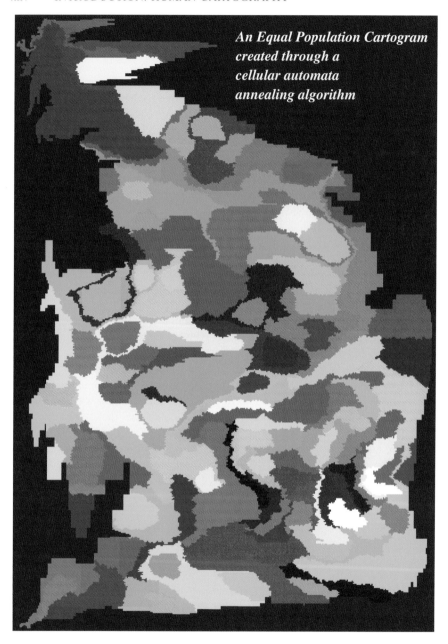

Figure I.6   The different coloured areas here represent some 250 counties and major cities (with populations over 50 000), which have been resized and reshaped in proportion to their 1981 census populations, keeping the rough shape of the coastline and the length of internal boundaries to a minimum. This continuous area cartogram was created on a 32K BBC microcomputer.

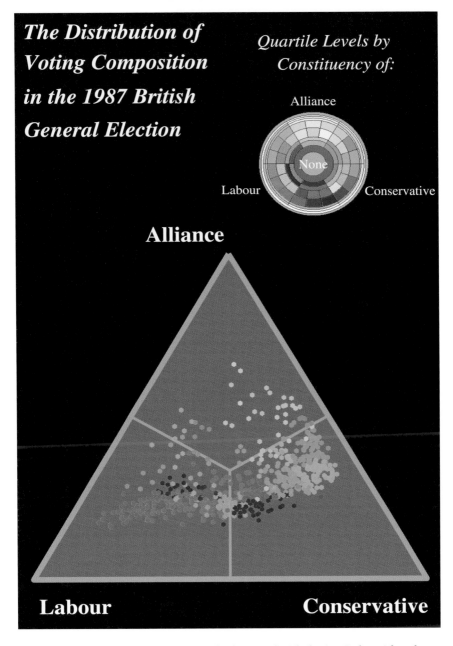

Figure I.7   In 1987 the Liberal Party had merged with the (ex-Labour) breakaway Social Democrat Party to form what was called the 'Alliance'. It won relatively few seats; seats are represented by all those hexagons shown in the top internal diamond of this electoral triangle, each coloured by the voting mix within the seat. Labour wins are bottom left. Conservative victories are bottom right.

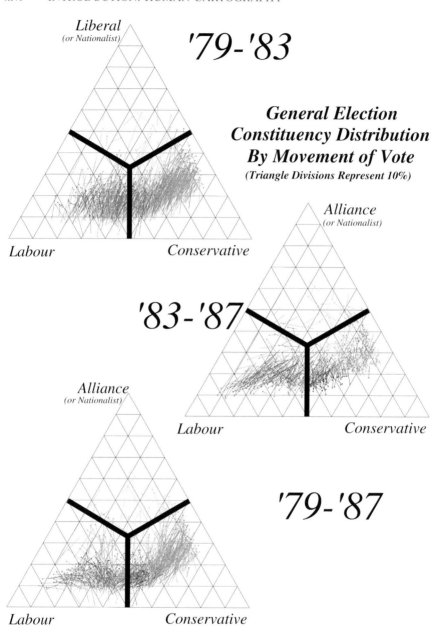

Figure I.8   On the electoral triangle each constituency is positioned by its three-way share of the vote. An arrow can be drawn to where that share moved to, coloured by the mix of votes at that political destination. The lower triangle, by combining two elections, shows most clearly how the country divided, the left-most arrows pointing leftwards, the right pointing right and up.

The methods employed here should hopefully encourage others to develop abstract imaging further. The ways people move about, day to day and year to year, is visualized in this book as streams flowing through space. Towards the end of the book are revealed images to depict the little that is known about large numbers of people, which are totally different from anything that would be recognised in current practice. A notional surface is proposed where the distance between points is made equal to how long it would take humans to travel between them, upon which we can then drape other distributions.

If enough researchers are inspired to experiment with the kinds of techniques shown here it may soon become possible to create true spacetime volumes of pattern and colour to depict the entire evolution of a single phenomenon, for example unemployment at every place, every month (Figure I.9). The alternative is to cut through this distribution, collapsing all of space to one point, to draw graphs of change over time.

Presented here is a methodology for studying relatively data-rich spatiotemporal distributions and their interrelations that is at odds with conventional approaches.[9] The alternative starting point of this work is to ask how you amalgamate individuals rather than subdivide society.

A logical unit of analysis does exist for the study of *spacetime* in human geography – it is a human life. As yet we have very little easily accessed information on individual people. However, of the whole of the population of Britain at least, census data are given at a resolution whereby, for national pictures, what is produced here using over 120 000 pixels would appear little different from pictures drawn with the benefit of such information – images using some 60 000 000 pixels, one pixel per person on a printed page. The images shown here are a little like those seen when everyone in a packed football stand holds a coloured card above their head to form part of a giant picture. The image can still appear as the full picture would, but somewhat blurred. Giving everyone a card of the same size does not distort the image.

Social science does need maps, but the maps that are currently drawn in its name, apart from often being bad examples of physical geography's cartography, are often bad social science. They make concentrations appear where they are not and dissolve existing patterns. They rarely portray anything but the most simple of spatial distributions, certainly not spacetime social evolution or the interrelation of a dozen different influences.

In this book some suggestions for new forms of visualization are given. When these images were first drawn, access to computers and to the software

---

[9] 'Eavesdropping in the conference bar, the cartographer's chatter is of the virtuoso Macintosh rather than the question of why and what we map. Are the mechanics of the new technology so preoccupying that cartographers have lost interest in the meaning of what they represent? And in its social consequences? And in the evidence that maps themselves can be said to embody a social structure? If material efficiency is allowed to dominate the design and construction of maps, we can see why the ethical issues tend to pass unnoticed. The technology of Geographic Information Systems (GIS) becomes the message, not just the new form or medium of our knowledge' (Harley, 1990, p. 7).

The Space/Time Trend of Unemployment in Britain, 1978-1990.

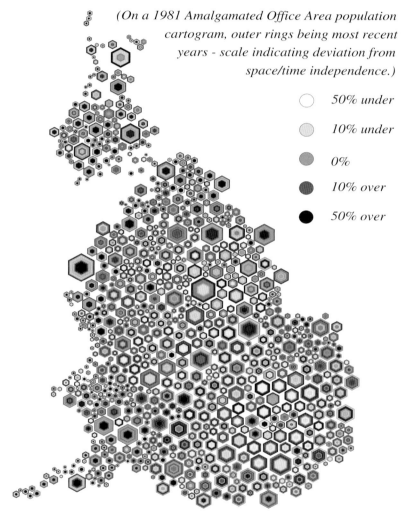

(On a 1981 Amalgamated Office Area population cartogram, outer rings being most recent years - scale indicating deviation from space/time independence.)

○ *50% under*

◐ *10% under*

● *0%*

● *10% over*

● *50% over*

Figure I.9   This is an image of all the publically available information at that time of unemployment rates, shown relative to the national average rate for each year. The sparkling picture shows the great variation from place to place with their relatively good and bad years, but cannot show the overall trend at the same time. A place that always bucked the trend would not stand out, which is most odd.

needed were far too restricted for there to be much chance of wider adoption. The images included here are only now being properly printed for the first time, despite having been drawn two decades ago.

In reproducing previously unseen old images in an updated context and with a revised text, it is hoped that through the subjects covered an alternative picture of Britain's recent past will develop in the reader's imagination. In this way the value of drawing images showing how everyone's lives are together arranged may become more apparent, not through these words and this writer's declarations, but simply through what you can begin to see by beginning to visualize the spatial social structure of Britain.

1

Envisioning information

We must create a new language, consider a transitory state of new illusions and layers of validity and accept the possibility that there may be no language to describe ultimate reality, beyond the language of visions.

(Denes, 1979, p. 3)

1.1 Visual thinking

Envisioning means bringing into the condition of vision for the purposes of contemplation, making visible, to enable visualization. It is what this book practises. For at least a century we have known that envisioning is about giving information to people in a form that is better suited to all our minds.

This work does not concentrate on the mechanics of getting information into and out of the machine, but instead with how you get it out to people (Figures 1.1 and 1.2). To communicate spatial structure is hard without involving the sense of sight.[1] Language, along with music, the most sophisticated use of hearing, is an excellent means of conveying ideas and thoughts, but cannot present a large amount of information in a structured form at speed.[2] Neither can touch or our other senses.

---

[1] 'Visual displays of information encourage a diversity of individual viewer styles and rates of editing, personalizing, reasoning, and understanding. Unlike speech, visual displays are simultaneously a wideband and a perceiver-controllable channel' (Tufte, 1990, p. 31).

[2] 'Human visual perception is performed by the most complex structure of the known universe, the visual cortex, that contains at least 10^{10} neurons, where each neuron on average contains 10^4 synapses (gates)' (Papathomas and Julesz, 1988, p. 355).

The Visualization of Spatial Social Structure, First Edition. Daniel Dorling.
© 2012 John Wiley & Sons, Ltd. Published 2012 by John Wiley & Sons, Ltd.

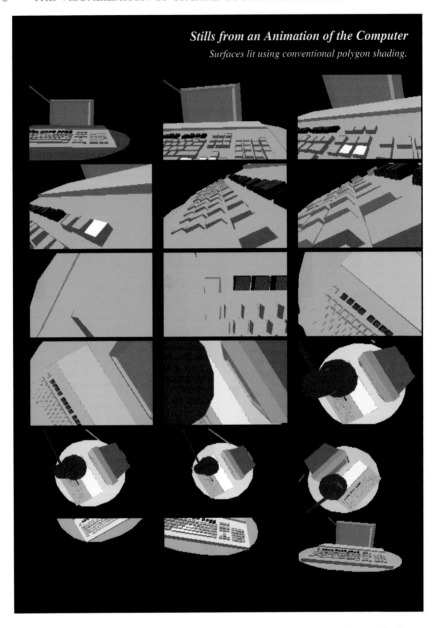

Figure 1.1   Software by Ace Computing was used to produce these still illustrations from a conventional video animation of the Acorn Archimedes computer. Polygon shading colours the entire surface of each facet a constant shade, which is determined by its colour and the light and shadow falling on it. The scenes produced are therefore crisp, but somewhat unrealistic. Note: this very early race tracing software produced images at even a low resolution very slowly and so parts of these 18 frames can appear jagged.

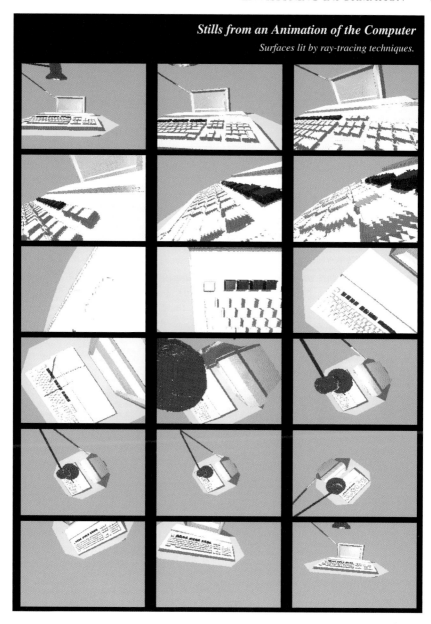

Stills from an Animation of the Computer

Surfaces lit by ray-tracing techniques.

Figure 1.2   Ray-tracing involves estimating the colour of every pixel in the image plane by calculating the trajectory of a ray of light on to the surface of a hypothetical object, where it can be reflected, diffracted or absorbed, and then through the screen to the eye of the viewer. During the late 1980s the algorithms were still in development. Later the advantages of this technique became more apparent. Note: the jagged nature of some of these images is accentuated when the virtual camera was very close to objects and it was then easy to over-expose the virtual 'film'.

When you look out of the window you can see a great deal in an instant.[3] The mind has an extremely powerful system for processing imagery that can instantly analyse a pattern of colours, of light and shade, and know that these are trees, houses or people. How long would it take to describe all that you can see in words?

Pictures alone are insufficient. This little book is only held together by its text. We have travelled a long way with our little symbols, the letters of the alphabet, which exist only because they were easy to scratch with a stick or form quickly with lips and tongue. Did our ancestors develop the most efficient means of communication or did they make do with what was possible?

The spatial structure of 1980s British society, which is envisaged in these pages, was made up of far more than a few large regions that can be named and divisions that could be measured. Social structure has a texture to it, a fine pattern, an elaborate organization, not unlike the fractal patterns to what were thought to be chaos, which were first revealed in the 1980s (Figure 1.3).

We depend on vision, we think visually, we talk in visual idioms and we dream in pictures, but we cannot easily turn a picture in our mind into something other people can see (and not everyone can see). An artist will take days to paint a single portrait. My parents' generation were the first to have easy access to the camera and mine were the first to receive the computer, which can turn a huge amount of data into pictures – snapshots of our society. In the future we may be able to speak visually and may be able to summon up an image to explain what we are trying to say. For now we still have to learn how.

1.2   Pictures over time

One of the great potentials of computer graphics is to provide a vision of what we might not otherwise be able to see in a photograph or real life.

(Dooley and Cohen, 1990, p. 307)

Our first permanent communications were cave paintings and our first textual scripts were made of pictures. Today the liquid-crystal display screen, which abounds with icons, is the modern cave wall (Figure 1.4); we have rushed forward to the beginnings of visual communication.[4]

---

[3] 'Humans can recognize unexpected objects in around 100 neuron-firing times' (Plantinga, 1988, p. 56).

[4] Although now the touch screen means the cave wall can react back. The subject matter of the earliest maps concerned people, the first map was a cave painting of figures dancing in a field. Later: 'Chinese literature tells us that maps were being used in the East as early as the 7th Century BC, while the earliest surviving examples of maps are clay tablets found at Nuzi, in northern Iraq. Believed to be from the period circa 2300 BC, they show rivers, settlements, land-holdings and hills' (Brannon, 1989, p. 38).

Ray-Traced Surfaces of the Mandelbrot and Julia Sets.

Grey shades of shadow and light.

Figure 1.3   Height of the surface shows the rate of divergence of each point on the complex plane, to infinity. The Mandelbrot and Julia sets essentially trace out one-dimensional lines in two-dimensional space. These are of such complexity, however, that a three-dimensional projection can be illuminating. These pictures are derived from simple equations. Note: the resolution reflects the original pixel sizes of the six screen shots are shown here.

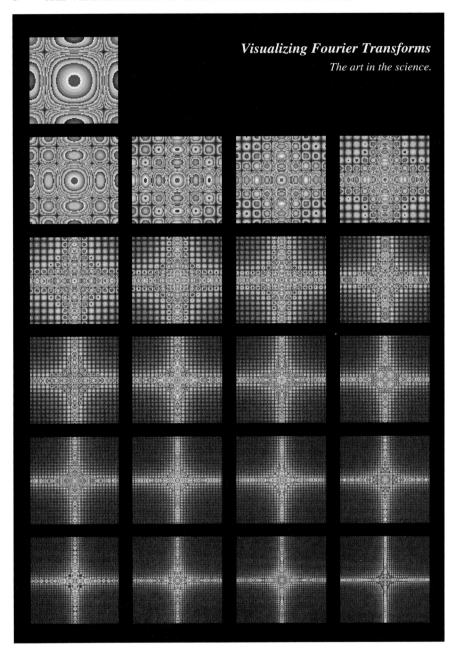

Figure 1.4   These images show the distribution of patterns of differing frequencies on the complex plane. The patterns show how science and art can merge. This is understandable when their objectives become less dissimilar. Different people might notice different patterns in the same pictures – there is some subjectivity in visualization because we have different experiences and expectations.

The first detailed maps were drawn on clay. They were invaluable objects for the control of territory or the projection of religious truth about the world. Maps were accumulations of innumerable stories, reams of parchment and hordes of figures, but they were also about power.

As map-making developed into the art of cartography, rules were formalized and conventions defined. Cartography is no longer a major discipline or even an important aspect of geography. Its modern tools can be used by children (Figure 1.5) and its conventions have been challenged as stale. It may currently be merging into a new, as yet unlabelled, discipline. This could be the discipline needed to collate knowledge on the graphical design of all that which now appears within touch-sensitive liquid-crystal displays, now that the displays interact. Disciplines change.

The nineteenth century saw the growth of an aversion, in science, to pictures. The graphs, which instruments traced on to paper, were immediately turned into supposedly more accurate and readable tables. Even in the early 1960s diagrams were said to be for people without mathematical imagination.[5] Nevertheless, statistical graphics did germinate in these surroundings.

The graph, bar chart and scatter diagram were invented and formalized. Rules for their construction were produced, while their supposed subservience to more advanced methods was made clear. By 1990 the cycle had come round again and a new breed of statistician appeared who saw visualization as paramount.

Computer graphics in the 1960s changed the picture. Swirling images were produced from simple formulae. It was immediately obvious that reading an equation told you little about what secrets it held. Before computer graphics, people were blind to the behaviour of relationships they thought they could easily understand (Figure 1.6). The programmers then turned their efforts to the possibilities of rendering abstract worlds.

Graphics have come in and out of favour in cycles through time. Their resurgences usually have more to do with taking advantage of new printing technologies and the availability of more abundant information than a basic understanding of their value.

Box 1.1 shows how computer graphics are constructed. At the time it was first drawn, being able to combine all these separate elements in a single 'graphics file' was novel. Creating such files required knowing that to place an island in a river within a park in a town in a county required the paths describing the boundaries of these objects to follow specific winding rules, to alternate from travelling clockwise to anticlockwise to clockwise again as you move outwards.

Today such graphics are routine on screens and ubiquitous on hand-held devices, but hardly anyone who uses them knows about things such as the

---

[5] 'In mathematics, it is considered the most flagrant gaucherie to use a diagram. "Graphics" is thought to be an inflated title for "mechanical drawing". In fact, all the intrinsically visible subjects; geography, graphics, and geometry, are suspected of being really grade school subjects, fit only for brains that are still undergoing biological maturation and whose harmfully misleading approach will have to be undone later' (Bunge, 1968, pp. 31–32).

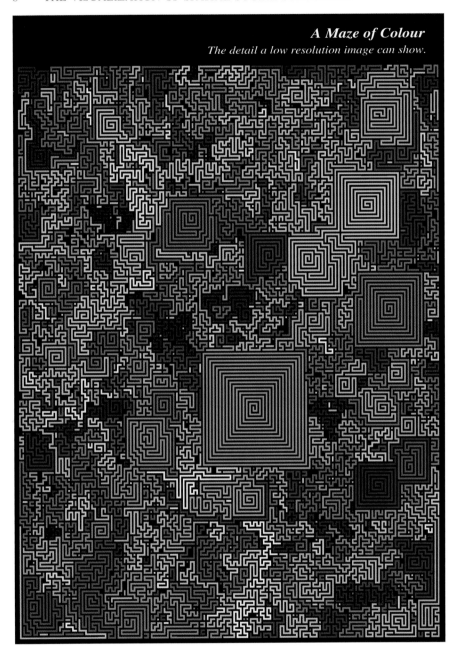

Figure 1.5   The maze is created by repeatedly choosing a starting position and colour at random, and then reversing direction, again at random, avoiding any obstacles. Great detail can be seen in this image, which consists of only 320 by 250 pixels. Some of the most detailed pictures shown in this book are made up of just over one million pixels. Often, though, only eight colours are used.

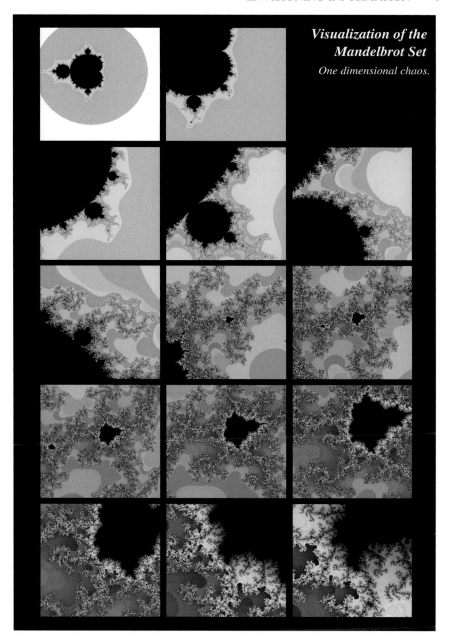

Figure 1.6   The same point in the plane is being repeatedly magnified in this visual series. By the end of the series the image is a million times larger in resolution. At this point the arithmetic accuracy of the computer used begins to fail. We can naturally appreciate complex images. There is no need to smooth out the beauty and diversity of reality either.

Box 1.1   Creating the graphics

Acorn drawfiles were used to create the illustrations in this book.

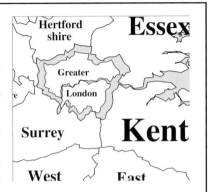

In 1989 a library of procedures was written specifically to produce these illustrations from data files.

Drawfiles are a sophisticated type of computer record. The record contains a list of objects, which can themselves be a list of objects.

Object can include **relationships** (with other objects), **information** (data from other files) and:

text – of a particular font, size, style and colour;

sprites – a pixelmap image (raster graphics);

paths – lines, curves and shapes (vector graphics).

In the example above the Greater London 'object' has been shrunk. In the drawfile it is tagged with its identification as County no. 1 and the relevant boundary date (1981). Making up the group is its perimeter, the river Thames and any islands in the river. All aspects of scaling, appropriate placement and hyphenation of names and colouring are automated. This automation was achieved by splitting names before parts of words such as 'shire' and scaling label font size to the boundary box of each area drawn on the map. Any feature of an object or group of objects can then be edited – interactively on the screen – as has been done here.

Once a drawfile representing a particular geography has been created, it can be transformed and additional information incorporated. For example, the places could be represented by faces instead of polygons, re-coloured and then merged with another drawfile.

'winding-rules' required to render complex topology. The rules have become embedded in machine code, the hidden instructions that make the computer work. As a result many more people can use computers, but a much lower proportion of those who use them can alter what it is the computer does as compared to the many programmers who could in the past. This lacuna was problematic for the development of visualization in the 1990s but, as software improved, what once had to be programmed became easier to create.

1.3   Beyond illustration

Visualization is a method of computing. It transforms the symbolic into the geometric, enabling researchers to observe their simulations and computations. Visualization offers a method of seeing the unseen.
(McCormick, DeFanti and Brown, 1987, p. 3)

Visualization is now a way of working – a methodology as much as a process. Not only does it differ from the use of script and figures – reading and calculating to understand – but also from conventional graphics, which aim to illustrate. Illustration is used to convey a discovery from one person to another, a discovery that was usually found by other means. Visualization is the transformation of numbers into pictures in order to see what a mass of figures cannot tell us, let alone could not inform others about (Figure 1.7).

Visualization, its early advocates suggested, is how discovery is made. For a time the method became the message.[6] Most visualization research today relies on huge quantities of numerical information.[7] Before you have such information, you can only write about what you think is happening. However, the problem for positivists (people who like countable things) is that once you have counted what is happening – who does what, who has what – how do you understand it?

How should we analyse information? Without visualization, statistical analysis gives you single figures, averages, correlations, parameters of assumed relationships, probabilities and so on. Such numbers are only of use if you know exactly what you want, but knowing what questions to ask is much harder than finding the answers to questions set. Social science is not about defining and testing simple hypotheses; it is about understanding societies.

There are many ways to begin studying society. All involve some form of ordering, of which the spatial is the most common. The patterns that visualizing society reveals usually turn out to show complex and subtle relationships that tax our mental capabilities to comprehend and explain. This is not a bad thing – stretching the mind forces the imagination. Hundreds of thousands of

---

[6] 'Computer graphics and image processing are technologies. Visualization, a term used in the industry since the 1987 publication of the National Science Foundation report "Visualization in Scientific Computing", represents much more than that. Visualization is a form of communication that transcends application and technological boundaries' (DeFanti, Brown and McCormick, 1989, p. 12).

[7] In the social sciences this was traditionally provided by voting data: 'In many ways elections are a positivist's dream. Millions of people go through the process of voting in numerous countries every year and these decisions are put together and published by areal units ready for analysis by social scientists' (Taylor, 1978, p. 153).

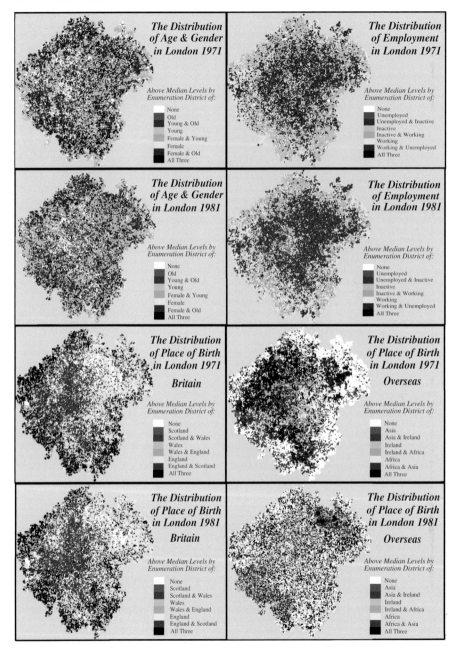

Figure 1.7   Each one of London's thousands of enumeration districts is shaded one of eight colours on each of these eight maps according to whether there are over-average or under-average numbers of old, young, male, female, working, unemployed or inactive people working in each, and by birthplace. Note how rapidly overseas birthplace distributions diffused during the 1970s.

digits are turned into a single picture. In terms of storage, most of the pictures in this book required more disk space individually than the entire (typeset) text.[8]

Illustration is to clarify – to make clear, pure or transparent. Visualization does not aim to see through the data; it aims to see into it. Methodology is about transforming reality to fit particular conceptions. The more we simplify, the more reality is blurred. Turning people and the events of their lives into numbers is bad enough. Throwing away almost all of those numbers is worse, and yet this is what we must do, in one elaborate form or another, if we are to try and understand without images.

1.4   Texture and colour

> *Colour is most useful, after position, to show information ... Colour deserves more attention than the others, especially in view of the hope for synthesis.*
>
> *(Tukey and Tukey, 1981, p. 193)*

If we are to envisage information we must first know what can be seen as well as what there is to see. To decide how to turn numbers into pictures we must know what pictures can contain and what is seen in them. The simplest pictures are constructed of pure black and white from basic geometrical shapes. What they contain, what the eye searches for, is pattern – from order, repetition, grouping and texture.[9] What the eye then does is to find breaks in that order and discover inconsistencies while ignoring irrelevancy. The eye does this because that is what it evolved to do, and to do so extremely quickly. In our minds we then compare what we see with what we have seen before; we learn to do that but evolved to be able to learn.

The eyes are constantly engaged in focusing, panning and zooming. They compare different sections of the image and home in on interesting detail (the eyes are designed to scan continuously – they cannot focus for long on a fixed point). The resolution of the eyes is enormous, but far finer at the point on which they are centred. This action can be enhanced when pictures are electronically produced, and can be instantaneously enlarged or reduced.

Research has provided explanations for some of the mechanisms through which vision may operate and suggests that it is easier for people to compare objects horizontally rather than vertically.[10] It also suggests that colour is an invaluable embellishment to basic vision (Box 1.2). It is wrong to think of it either as adding another dimension or merely supplying some further minor tagging of data to existing features of the graphic. It alters the character of the image.

---

[8] Even the most complex image shown here could be fitted on a single 1.4 Mb floppy disk!

[9] See Bachi (1968), Hunt (1968), Tobler (1973b) and Levkowitz (1988, p. v).

[10] '... because the eyes are spread apart horizontally – as is, presumably, the spatial medium they feed – they have a greater horizontal scope' (Kosslyn, 1983, p. 71).

Box 1.2   Printing in colour

The device first used to print the colour illustrations in this book was a ColorView 5912 plotter printer manufactured by CalComp in 1988. The plotter could produce pixels of eight colours by overlaying sheets of magenta, cyan and yellow film with an A4 resolution of 2048 by 1600 pixels. A greater range of colour was possible by using dithered patterns of the eight colours actually available. On-screen text could be more satisfactorily produced through the use of anti-aliasing techniques built into the computer software.

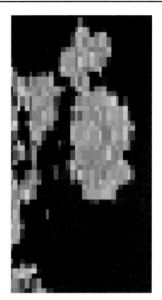

A driver was specifically written to convert red-green-blue output from the screen to the magenta-cyan-yellow form suitable for printing.

Each colour print, originally A4 size, could have over three million individual *bits* of information, and took half an hour to print.

Different colours are perceived variably and convey loaded meanings on their own, even more so in combination. The human eye is poor at focusing on blue. Red and green do not combine to form reddish-green. Colour adds another level, but not dimension, of complexity. The careful use of colour can convey more of the depth of spatial organisation. In particular, when used in bivariate and trivariate mapping, colour can show how several variables are related to each other on a single map, although the keys are complex (Figure 1.8).

Trivariate mapping is both a contentious and potentially highly effective technique.[11] It has been suggested that the printer's primary triplet of cyan, yellow and magenta be employed[12] (or the computer's red, blue and

---

[11] 'It is far more difficult to distinguish the amounts of the three primary colors painted simultaneously onto a point in space, but it is possible (barely possible) to do so. Therefore a crude, but effective, way exists for displaying three functions of three independent variables' (Staudhammer, 1975, p. 183).

[12] '. . . maps with the same scale can be superimposed three by three. It is sufficient to transcribe them on three different color films: cyan-blue, yellow, magenta-red' (Bertin, 1981, p. 163).

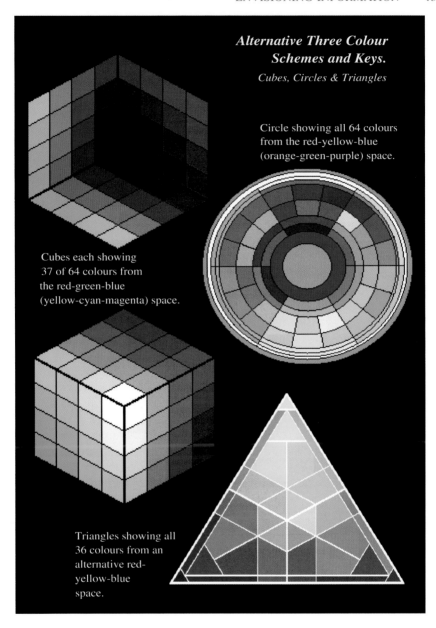

Alternative Three Colour
Schemes and Keys.
Cubes, Circles & Triangles

Circle showing all 64 colours
from the red-yellow-blue
(orange-green-purple) space.

Cubes each showing
37 of 64 colours from
the red-green-blue
(yellow-cyan-magenta) space.

Triangles showing all
36 colours from an
alternative red-
yellow-blue
space.

Figure 1.8   Several of the visual techniques that were used to define the various three colour schemes in this work are shown. The triangle was eventually deemed the most successful. However, as in a triangle, the three quantities must sum to unity; it is only showing a two-and-a-half-dimensional range. The colours here vary slightly from those intended and produced by the original printer used. Note: these four images were originally produced as a bit-map of pixels of colour not as a vector graphics files and so are pixilated.

green)[13]. However, in this book I suggest that the most intuitively appealing combination is the painter's red, blue and yellow – which fortunately also coincide with Britain's major political parties' symbols and also with national colours, and have many other useful associations.

1.5   Perspective and detail

> *Generalization, if you wish to call it that, occurs spontaneously in all perception. Complex though a map may be, the mind derives from it a simplified pattern.*
>
> *(Arnheim, 1976, p. 9)*

The most powerful ability of the eye–brain working in combination is generalization.[14] The brain only ever sees through the constantly changing light intensities, which are measured by the retina. These are analysed by the brain to allow instant assumptions to be made, before more careful inspection is undertaken (Figures 1.9, 1.10 and 1.11). Such ability is essential to our survival in everyday life; it was even more so in the past. Through visualization we are utilizing one of the most finely tuned pieces of evolutionary good fortune.

We live in a three-dimensional world, despite having as near to two-dimensional vision as often makes little difference. Perspective is the name given to the effect of projecting a three-dimensional scene on to our two-dimensional retinas; we use it to try to reconstruct three-dimensional form. Although we do have binocular vision, if you close one eye you lose little feel for the three-dimensional reality. For the most part we move about in two dimensions and, in fact, have a far weaker grasp of the real three-dimensional world than we may imagine (as illustrated here in Chapter 9).

It is often claimed that expensive equipment which allows volumes (commercially known as '3D') to be created and seen is at the forefront of visualization. Stereoscopic vision, though, might not be as great an asset to visualization as it is often thought to be in seeing the real world. Stereo vision works well at gauging position when nothing is moving behind or in front of anything else. Once things begin to move, though, it can become a confusing irrelevancy. In visualization, if we want things to move, then it is usually through animation that they move.

Animation can be used for much more than understanding three-dimensional form. As the creation of a changing or moving image, it can add another level of sophistication to two-dimensional visualization. However, like colour, animation (employing time) is not the same type of dimension as the spatial. In animation

---

[13] 'It is possible to express a trivariate distribution by mapping each variable onto one of the dimensions of (red, green and blue) color space' (Sibert J.L. 1980, p. 214).

[14] See Tobler (1968, 1969, 1989b), Rhind (1975c), Lavin (1986) and Herzog (1989).

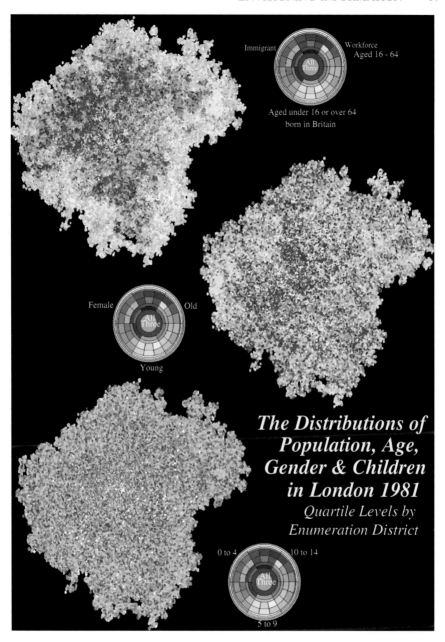

Figure 1.9   Showing three equal population projections of the 16 975 enumeration districts of the 1981 census in London. Here the scales are set to highlight the distributions within the Capital. The centre is dominated by people who have migrated in, by the old and by slightly more of the youngest children, before they move outwards. The images show these are generalisations.

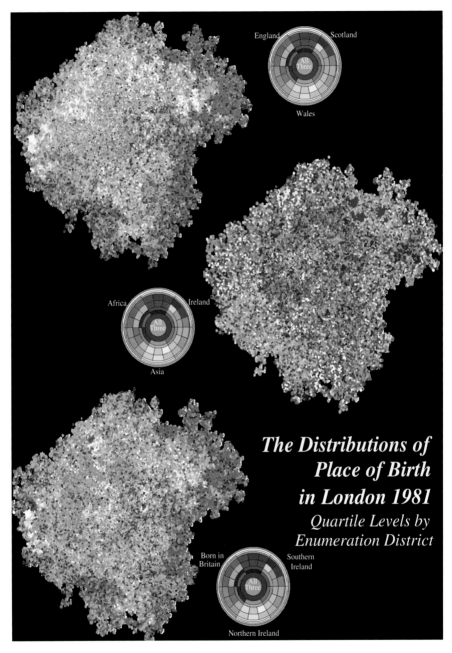

Figure 1.10   The Scots and Welsh dominated the West end of the Capital in unusual numbers, while those from other countries were, by 1981, more spread to the East. The Irish immigrants tended to fill the holes left between the patterns for other nations. North and South of the Irish border are differentiated here to illustrate how they were subtly differently distributed in London then.

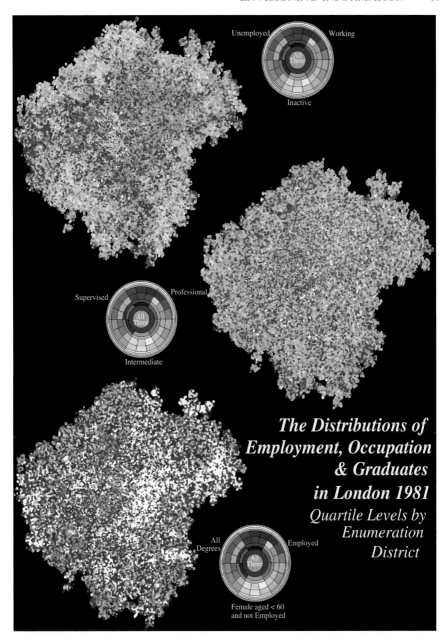

Figure 1.11   Notice how these three sets of three distributions mirror each other. A serpentine streak of affluence snakes through the centre of the City, closely flanked by areas of the highest unemployment and lowest job status. In the bottom picture, white areas are where all three categories were in the lowest quartile. More had degrees where women under 60 did not work.

things must change smoothly and relatively slowly.[15] If objects change their colour it can confuse; if too much is happening we will not have enough time to comprehend it.

Surprisingly, animation takes us back towards illustration. It requires simplicity to work. Far more useful are interactive graphics – moving pictures that the viewer controls. An App on a mobile phone is an easy contemporary example of an interactive graphic. The viewer has control. This is not only control over how fast or slow or where the pictures move, but simultaneously over what they contain and how they are presented. This interactivity is a possible next step in visualization that may allow more of the types of image shown here to be created, understood, manipulated and used.

1.6   Pattern and illusion

> ... the differences between maps and other forms of graphic information are not as great as they appear. All types of graphic information are different solutions to a common problem: our limited capacity to remember unprocessed information. By removing the limitations of short term memory, graphic information allows us to do kinds of thinking which are difficult or impossible in other ways.
>
> (Phillips, 1989, p. 25)

We do not think in a three-dimensional geometry – many tests have shown this, although a variation in propensity for such thinking may exist between men and women.[16] The geometry of visual thinking is essentially two-dimensional. We also have a poor visual memory; we remember what we extract from images rather than the images themselves. Furthermore, the emotional overtones of colour are perceived differently by different people.

The colour blind cannot see the full trivariate range. Advocating trivariate mapping went against many of the embryonic tenets of visualization, but it is questionable how much they were guided by what was possible, rather than what was desirable. Why use the illusion of three dimensions if it adds so little information to an image while causing so much confusion? Perspective views

---

[15] In a visualization, involving animation is often best if objects don't move very much: 'Several trial films revealed one very necessary characteristic of animated mapping: simplicity and extreme clarity are essential. In a static map, the reader has time to interpret complex or unclear information. However, this is not the case in animated mapping where the image must be interpreted immediately' (Mounsey, 1982, p. 130).

[16] The kind of test that uncovers our general inability to think in 3D geometry is to try to imagine the shape made by a hot cube pressed, point down, into a think sheet of ice (Parslow, 1987). The answer is an equilateral triangle (if the cube is held steady). Many people say 'square'. Men are, apparently, on average a little better at such imagining than women. It is one of the very few tests where men do perform better. Researchers speculate that it helps in throwing spears better, but also that throwing spears may have been more important for impressing women than for catching meat.

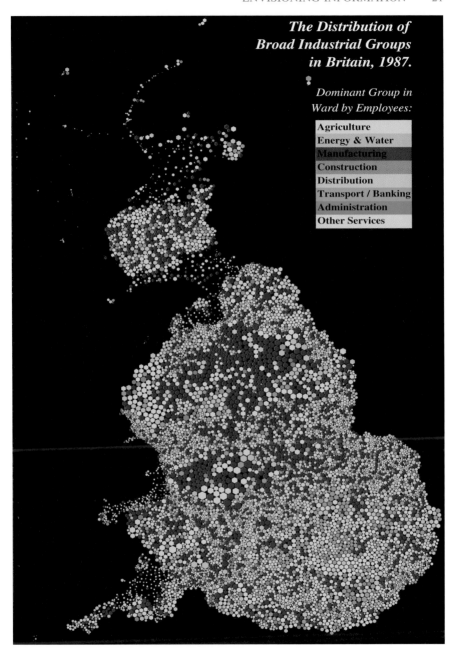

The Distribution of
Broad Industrial Groups
in Britain, 1987.

Dominant Group in
Ward by Employees:

Agriculture
Energy & Water
Manufacturing
Construction
Distribution
Transport / Banking
Administration
Other Services

Figure 1.12   The 1987 census of employment by 10 444 frozen census wards on the equal population cartogram. Manufacturing industries were more dominant in Northern England and service industries more prevalent in the South, but both were mixed about the other. The coalfields of Yorkshire can be clearly made out (Energy), while agricultural workers are visible in smaller rural wards.

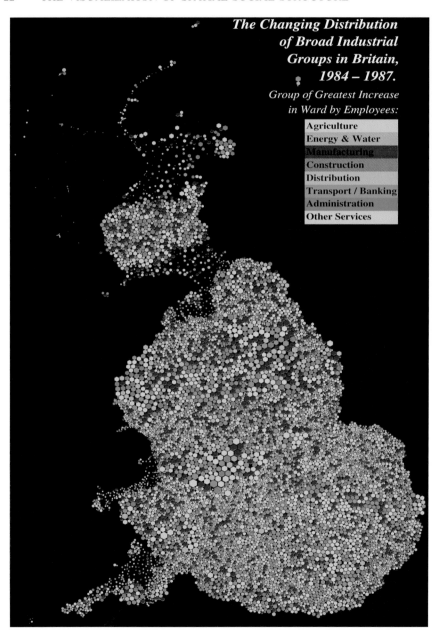

Figure 1.13   The changes show no clear pattern, although agriculture may have increased a little near many large towns. The blue and green of growing service industries became more common in the Capital, but expanding industries were scattered everywhere. This is only the change in three years and the industrial landscape tends to change more slowly than that.

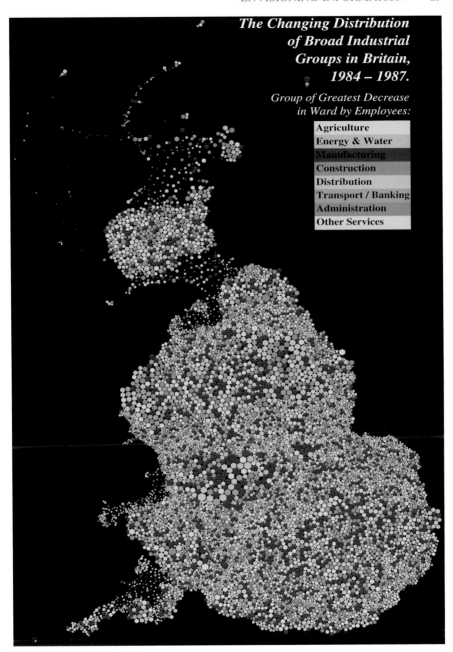

The Changing Distribution of Broad Industrial Groups in Britain, 1984 – 1987.

Group of Greatest Decrease in Ward by Employees:

Agriculture
Energy & Water
Manufacturing
Construction
Distribution
Transport / Banking
Administration
Other Services

Figure 1.14   The most noticeable pattern here is of the mining jobs, which were lost after the national miner's strike of 1984. They were lost in greatest numbers in Yorkshire, where support for the strike had been strongest. The National Coal Board was the biggest employer in many towns in that county. Manufacturing declined much more than service industries.

are pretty, but not especially useful unless it is three-dimensional geometry in which you are interested.

Animation, like perspective viewing, is also not as invaluable as has been claimed. You cannot hold a moving picture in your mind as well as you can hold a static image, and comparison of two dynamics, of two animations, is very difficult.

Animation can tell a story. Visualization, more often, allows you to find a story to tell (Figures 1.12, 1.13 and 1.14). Much more importantly, with both animation and perspective views, you are limited to producing very simple pictures if you are to be able to understand them. Both perspective views and animation are included in this work and they produce nice illustrations, but until the viewer can easily control what is viewed, through interactive visualization, their utility is limited. Even if you are reading this on a Kindle (book reader) or on a different kind of computer screen, you (probably) cannot as yet spin the images in this book.

The use of colour greatly augments what can be seen in a two-dimensional image. However, use of colour is still expensive in publishing, even in 2012. Duplicating the prints shown here was almost impossible 21 years ago.[17] Colour can also invoke unintended ideas: good and bad, hot and cold, near and distant hues – but these can be used intentionally too.

In this book colour is generally not included to make the pictures prettier. I have used colour to include extra information in the image and to show how to display more complex data. Often it is added as a final embellishment to elaborate on how all is connected as other facets of the social structure are added as further elements in an image (Figure 1.15). Sometimes removing colour clarifies (Figure 1.16).

1.7   From mind to mind

That very ancient merger of Geography, Geometry and Graphics still exists and, if anything, with increasing vitality. Many breakthroughs still lie ahead. The map is the geographer's laboratory.

(Warntz, 1973, p. 85)

The argument in this chapter has developed from the initial desire to allow people to convey what is in their mind, in a form others can see, to the point where individuals are able to see and paint their own information. If we had all been mute and suddenly were able, with the aid of a machine, to make sounds,

---

[17] It may shock people today to read how expensive colour used to be, even more expensive than computers used to be: 'Currently, however, the cost of publishing two-color plates in some scientific journals represents more than half the cost of the laboratory computer that controls the experiment, stores the data, and displays the results. Therefore, the use of color figures (which can best present the results) might be hard to justify' (Long, Lyons and Lam, 1990, p. 138).

Box 1.3 Recording the places

All information about places, concise enough to be edited manually, was stored in 'Comma Separated Value' (CSV) files. These files can be read by many applications on different computer systems, in particular by spreadsheets – allowing complex manipulation to be easily accomplished. An example of the beginning of a CSV file containing information used to create a drawfile of counties is:

```
"$.GIS.Area.Ward.County.Sheet", 1,64,4
"County Topology and Statistics"
"Number", "Name", "Residents", "Neighbours", "Neighbour"
"1981"
1, "Greater London", 6713130,6,0,29,22,43,26,11
29, "Kent", 1467079,5,0,1,21,43,22
22, "Essex", 1474126,6,0,12,42,1,26,29
43, "Surrey", 1004332,8,21,45,1,29,0,24,10,11
26, "Essex", 1474126,6,0,12,42,1,26,29
11, "Buckinghamshire", 567979,7,43,26,34,9,38,1,10 ...
```

The first line gives the filename, number of tables, number of areal units and fixed variables. The second line describes the file, the third has variable names and the fourth holds temporal information. This header is followed by the relevant numbers and text for each place. Notice that the records can be in any order and of variable length. They are easily edited individually as this is a text file.

A library of procedures was written to manipulate these files, in particular to allow any other application to read and write to them, taking advantage of an interpreted language, which allowed the procedures themselves to invoke routines from the applications that had called them.

what sounds should we make? In the past we made sounds by knocking sticks together, so we get the machines to imitate those noises (drum machines). But surely, we think, is there more? According to Aristotle, 'thought is impossible without an image'.[18]

Our vision has a much higher bandwidth than our hearing. We can see thousands of stars, watch sunsets, view landscapes and survey half a million people in a crowd. Naturally we begin to paint things by getting them to look like recognisable objects, chaotic functions to look like mountain ranges or an island archipelago, flowing energy to appear as running water. In this work pictures are often based on natural things that have two-dimensional structures, ranging

---

[18] 'Memory, even the memory of concepts, does not take place without an image' (Kosslyn, 1983, p. 5).

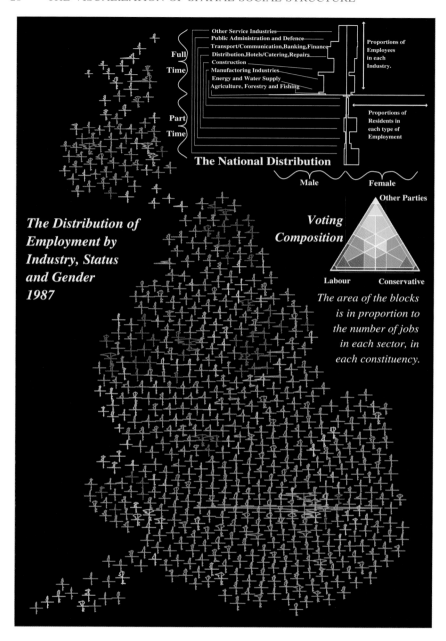

Figure 1.15   The 1987 census of employment figures for 633 parliamentary constituencies on an equal population cartogram, coloured by the 1987 general election results. Each glyph shows the share of full-time employment by industry above the line, part-time below the line, male employment to the left, female to the right. The area within the glyph is proportional to the number of jobs.

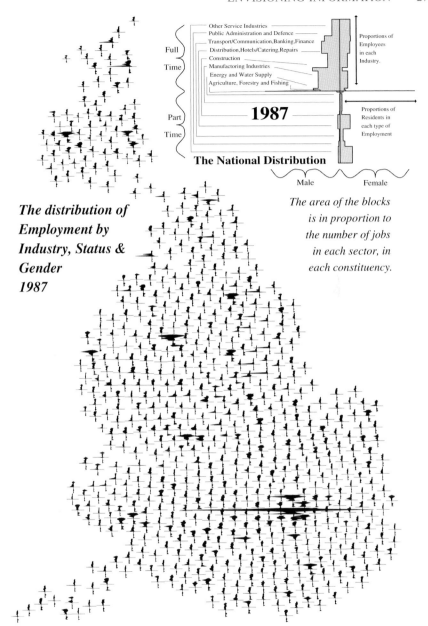

The distribution of Employment by Industry, Status & Gender 1987

The area of the blocks is in proportion to the number of jobs in each sector, in each constituency.

Figure 1.16   When the voting pattern is taken off this image and just the industrial composition of employment is shown other features become apparent. Regional service centres are clear with their mushroom-like structure. Although change is not shown here those areas that gained jobs did not necessarily swing to the Conservative Party who were in power and those that lost employment did not always turn towards the Labour Party.

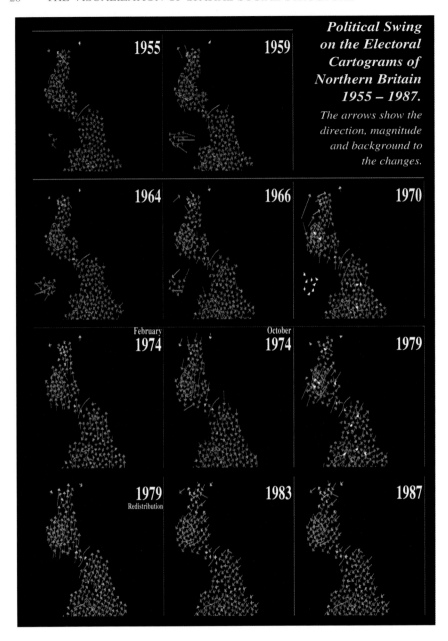

Figure 1.17   This series of images all show the Northern half of the 705 consistent parliamentary constituencies on the equal population cartogram. In the North the picture is dominated by the Labour Party. These are key frames taken from a video that shows how the vote swung to make Scotland the major Labour stronghold and took Northern Ireland away from the Conservative Party.

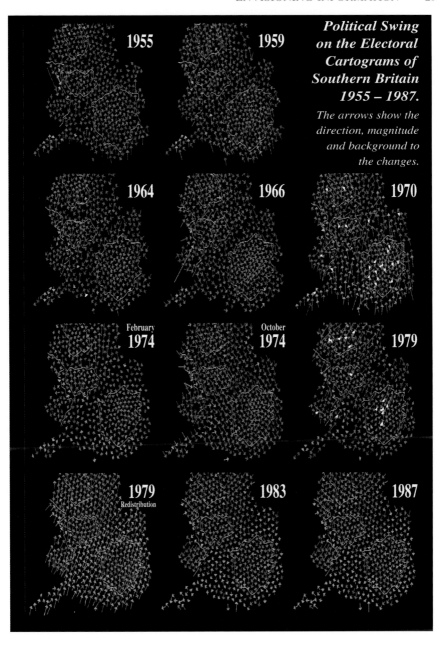

Figure 1.18   Southern Britain: the major conurbation boundaries can just be made out in the graphics. In the animation, to which these are the key frames, the arrows can be seen swivelling around between elections like so many synchronised swimmers, in near unison. The consolidation at the end of the period, as groups of places begin to move in different directions, is particularly well marked.

from honeycombs and cobwebs to crowds of upturned faces and flocks of arrows (Figures 1.17 and 1.18).

Here visualization is used to make millions of figures understandable without massacring their meaning, without reducing them to tables, graphs, crude maps or models. Visualization is not about simplifying; it is about revealing, through the process of aiding understanding. If we are to understand the structure of society we must find ways of imagining it. This book demonstrates how large amounts of simple information can be shown and then goes on to increase the potential of the graphics by conveying increasingly detailed information (Box 1.3).

The pictures shown here are of things that cannot be easily (or adequately) described, discussed or modelled, and yet many people who see them expect to understand them in an instant, even when they may fail to understand the long complicated narratives, which explain them badly, or the intricate mathematical models, which could represent them inadequately.[19]

If you want to know the shape of Britain you look at a map. You can then go on to investigate rivers and mountains, lakes and bays. Here you can look at the shape of 1980s British society – not the physical landscape but the human one. However, to know the shape you have to look at the picture, you cannot just describe it in words.[20]

---

[19] It remains the case that we cannot produce images at will. 'The unhappy thing about all this, of course, is that whereas I have the ability (and we all have the ability if we're sighted) to take images in at a fantastic rate, I have no ability to create images with the same facility. This is a one-way street. On the other hand, I can create language and symbols at about the same rate I can take them in, which means I can create speech at about the same pace that I can listen to it. So it is not at all unexpected that for most of us language seems to be the main carrier of our thoughts because that is the thing we can hear ourselves saying and were conscious of its use' (Huggins, 1973, p. 37).

[20] One day soon you may be able to create images as quickly as words, but at that point a new form of language will have been created. The last time humans invented a new form of language was when they became human over 60 000 years ago. It can take as long as 21 years to draw an image and then have it printed (as in this book). Try to imagine if that process took 0.21 of a second and then you showed me how you imagined it too. Do this and we have a visual conversation; we change the pictures in our minds as we draw them for each other, agreeing, disagreeing and learning.

2

People, spaces and places

The practices through which social structure is both expressed and reproduced cannot be divorced from the structuring of space and the use of spatial structures. Previously structured space both constrains and enables the reproduction of social practices and social structure.

The social becomes the spatial.

The spatial becomes the social.

(Pred, 1986, p. 198)

2.1   Which people?

I am interested in the lives of the people of Britain over two decades, the 1970s and the 1980s. This is because I am one of the people who was alive then. By 1990 I had been counted as a birth in 1968, a child in two censuses, as a migrant by the Health Service, as a claimant of unemployment benefit, as a voter in a general election.[1] As I have been counted, so have been millions of others. Surely, I thought, all these numbers can be turned into a picture of the people in the country in which I grew up?

What is it about these people that we wish to understand? Trivially, it is who these people were, how old they were, what they did. Fundamentally, we ask what the relationship between these people was and what was the structure of the society in which they lived? You cannot have grown up in Britain in these decades without having felt the weight of this structure, how it affected your life,

---

[1] I had only voted in one general election when I first wrote the words that you read heavily edited here. I have voted in another five since, and my children have been born and filled in their first and possibly last census form in 2011 (all population censuses after 2011 have been cancelled in Britain; it is claimed this will save money). My youngest son listed his ethnicity as 'purple' when he looked at the options, as that year he was set in the 'purple group' for reading at school!

The Visualization of Spatial Social Structure, First Edition. Daniel Dorling.

defined your opportunities and altered your destiny. When you felt it you may easily not have been aware that it was this you were feeling. It could just have felt normal to have been treated as you were and to learn to behave to others in particular ways. Looking back you know it was as much about the times and places you were in (Figure 2.1).

Whether you were male or female, where you went to school . . . you knew (or now know) it made a difference, but often you did not know quite what difference it made. More importantly, you could not necessarily easily have known the effect of the social structure on everybody else. It is the relationship between the gains of some people, and the losses of others, that would easily escape any single individual's perception. An individual may be able to imagine some of it, but not to *picture* all of it.

It is only by first seeing what you wish to comprehend, that you can begin to understand why and how it exists. Just to create an image of the simplest manifestations of the structure of society is a difficult undertaking (Box 2.1). The story that is told should treat everybody's part in it as equally important, as all their lives should, at least in the way that their history is told, have equal value (Figure 2.2).

British society was chosen as the subject of this work because of the practicalities of the exercise, since getting digital data in the 1980s was not easy. That the line chosen around Britain divides land from water is convenient, but not the reason for its imposition. The line around this country divides the experiences of most of its people from those outside.[2] The sea may present no great economic or political barrier any longer, especially with a tunnel now running under it, but it is still a very strong social and cultural divide.

Some of the sharpest divisions are the closest – those separating Northern Ireland from Britain, the United Kingdom from Europe – although they are much less in magnitude than inequalities further afield. Eventually it may be possible to undertake this kind of study across those lines, although there is still no social atlas of Europe to view, even today. But extension must not be used as another excuse for amalgamation – creating the average English, Scottish or Welsh man and woman (Figure 2.3). Before we can begin to understand world society better, it is prudent to delve into our own, to see just what we are comparing others with.[3]

---

[2] I agreed with this: 'It should be noted that Great Britain (England, Wales, and Scotland) is not the United Kingdom (which includes Northern Ireland as well). But, given the fact that Ireland as a whole constitutes a separate land mass, that it was historically governed as a colony of Great Britain, that the division of Northern and Southern Ireland occurred only in 1922, and that Northern Ireland itself contains but six counties – for these reasons, we restricted the study to the single land mass of Great Britain for which the requisite data were readily available' (Massey and Stephan, 1977, p. 352).

[3] In order to draw transformed maps of societies at a world level with colleagues we did at first amalgamate statistics to the UK level, but later we broke up the state again and drew maps of the population of the whole planet by 2010, which showed how many people lived in each small area. This resulted in a new world map projection upon which other patterns can be drawn (see Hennig *et al.*, 2010, pp. 66–69).

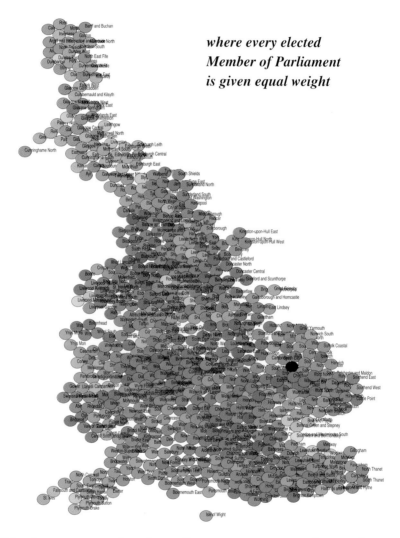

Parliamentary Constituencies in 1981

*where every elected
Member of Parliament
is given equal weight*

Figure 2.1   A prototype version of a parliamentary cartogram where each constituency is represented by an oval of equal size. Many ovals overlap because the cartogram is still in the process of being made. Each oval is coloured by the results of a general election using red-yellow-blue colour shading. The name of each constituency is included, starting from within its oval.

Box 2.1   Drawing the maps

For the mapping in this book a program was written, in 1989, to perform automated cartography. The program combined information from geometry and lookup data files with CSV text files describing the topology and other attributes of places, to produce the desired map as a drawfile, which could then be manipulated further. The shading and names of places was given in the text file and a set of simple rules applied to annotate the areas. Names were split at spaces, hyphens, commas, before the syllable 'shire' and wherever an underscore had been inserted. The text was then scaled to fit in a rectangle within the place and centred.

Places could be represented by complex 'paths' rather than simple polygons, so all the islands and lochs of Western Scotland were easily combined as a single list of the boundaries of Strathclyde. Paths could also overlap and the space covered did not have to be completely allocated to areal units, as it often does in traditional GIS systems. The places did not have to be given in any particular order as each was tagged in the drawfile with an identification code, permitting many other graphical possibilities, which will be shown later.

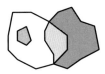

Creating the maps as drawfiles also meant that they could be produced on almost any printer available or dragged into documents such as this.

2.2   Why study places?

People live in different worlds even though they share the same locality: there is no single community or quarter. What is pleasantly 'old' for one person is decayed and broken for another.

(Wright, 1989, p. 290)

Place matters to the relationships between people (Figure 2.4). Just as your place in time so obviously constrains and determines your life, so your place in space limits and creates the possibilities in your world. It is not the actual position in space, as it is not the actual position in time that does this, but who shares that place, who shares your time.

However, times and places are fundamentally different things. As we live we must all share the same moment in time, but, in existing, must be spread over space into different places. This diffusion of settlement in space, juxtaposed with the concentration in time, defines the dimensions over which experiences

Parliamentary Constituency Area Population Cartogram

Annotated for identification.

Figure 2.2   On this cartogram every 1981 constituency is drawn with its area in proportion to its population and touching as many others that it originally neighboured as possible while touching as few as possible that it did not. Islands touch the areas they had ferry connections to. Names are added to aid understanding, which on a computer screen could have fitted inside each circle, visible by zooming.

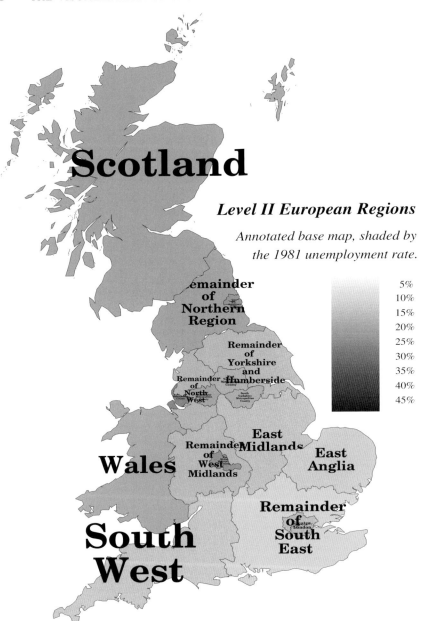

Scotland

Level II European Regions

*Annotated base map, shaded by
the 1981 unemployment rate.*

5%
10%
15%
20%
25%
30%
35%
40%
45%

emainder
of
Northern
Region

Remainder
of
Yorkshire
and
Remainder Humberside
of
North
West

East
Remainde Midlands
of
Wales   West
Midlands

East
Anglia

Remainder
Of
South
East

South
West

*Figure 2.3   The first of a series of maps showing unemployment from the 1981
census on equal land area maps with automatic annotation. The 17 areas here
were regions used by the European Community during the 1980s to present
coarse statistics about Britain and allocate regional grants. Above this level of
amalgamation Britain is often divided into 11 regions, or just England, Scotland
and Wales.*

Islands

Highland

Grampian

Tayside

Centr.. Fife

Strathclyde   Lothian

Counties and Scottish Regions

*Annotated base map, shaded by
the 1981 unemployment rate.*

Borders

Dumfries
and
Galloway

Northumberland

Tyne
and
.ar

Cumbria   Durham

5%
10%
15%
20%
25%
30%
35%
40%
45%

North
Yorkshire

Lancashire

West
Yorkshire   Humberside

Greater
Manchester   South
Yorkshire

Cheshire   Nottingham-
shire   Lincoln
shire

Gwynedd   Clwyd   erbyshir

Stafford-
shire   Leicester
shire   Norfolk

Shropshire   West
Midla

Powys   Warwick
shire   Northampton
shire   Cambridge
shire   Suffolk

Dyfed   Hereford
and
Worcester   Bedford
shire

Gloucester
shire   Oxford
shire   uckingham Hertford
shire   shire

West
Glamorgan   Mid
Glamorgan   Gwent   Greater
London   Essex

Avon   Berkshire

Wiltshire   Surrey   Kent

Somerset   Hampshire   West
Sussex   East
Sussex

Devon   Dorset   Isle
of
Wight

Cornwall

*Figure 2.4   Continuously shaded map for these 64 areas drawn from 1981 census
figures and projected on an equal land area map with annotation applied automatically. Counties are often used for mapping statistics. Councils often used to
be elected to administer government in these areas, from county towns. The major
metropolitan areas no longer exist as administrative units.*

can differ. The organisation of people over space, and through time, is what makes place important.

Constrained by the limits of time, people are forced to live close to where they work. In cities they work together, but live apart. We used to live together in villages and work apart in fields; now many live apart in suburbs and work together in offices. This spatial organisation reflects the need for people to live together and the wishes of some to be apart. As they are rewarded unequally at work (if they work), this inequality is reflected in where, and how, they live.[4]

In a country with growing inequalities more and more neighbourhoods are created as areas where most of the people have a comparable income. Here they live in houses that are alike, have similar backgrounds and, to some extent, a common future. When a firm closes down only those localities from which its employees came are directly affected. If the supply of labour is spatially compact, so too will be the spatial impact of job losses. Only pupils in the locality of a certain state school will go to that school. People downwind of a particular source of pollution will be most affected by it.

If people in one place suffer, so eventually will all others in some way. However, it is the spatial reinforcement of certain trends that makes the importance of place so clear. Just as the pattern of commuting allows the neighbourhoods to exist, so the pattern of migration serves to exacerbate neighbourhood differences (Figure 2.5). As a few people in poor neighbourhoods do well, they move to richer ones. More importantly, the vast majority of migration is between similar places in the spatial social hierarchy, reinforcing and perpetuating the existing differences.

The term 'locality' defines, here, a group of people who live in close proximity. They do not necessarily have to know or even recognise one another. What the study of human geography has shown is that they will tend to have more in common with each other than with outsiders.[5] This is because of what put them there, keeps them there and moves them away – the forces that sort people in space, the institutions underlying social structure.

Place is important to the understanding of the social structure of society because it is through places that the structure is most directly visible. Not only is it visible in our everyday lives, but some of its many facets can be made visible, if still blurrily, on paper (Box 2.2).

---

[4] Showing any of this is harder than describing it. A century and a half ago a man, 'the Superintendent', tried to work out a way of mapping the production of a given agricultural crop. He tried to do this using a topographic map and failed: 'the Superintendent satisfied himself that no one simple ratio could be found which would not, in many cases, grossly exaggerate, and in other cases as unjustly disparage, the importance of the crop to the county and the county to the crop' (Walker, 1870, p. 367).

[5] This is known as Tobler's first law: 'Everything is related to everything else, but near things are more related than distant things', http://en.wikipedia.org/wiki/Tobler's_first_law_of_geography, which perhaps should have said ' . . . are usually more related . . . '. More prosaically: 'To insist on the continuing importance of place, therefore, is not to deny that processes beyond the locality have become important determinants of what happens in places. But it is still in places that lives are lived, economic and symbolic interests are defined, information from local and extra-local sources is interpreted and takes on meaning, and political discussions are carried on' (Agnew, 1987, p. 2).

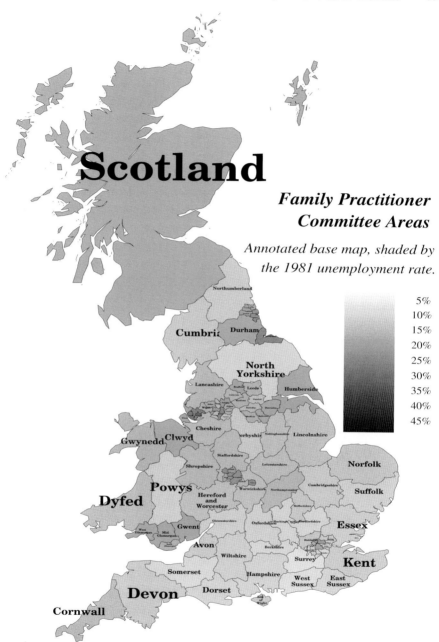

Scotland

*Family Practitioner
Committee Areas*

*Annotated base map, shaded by
the 1981 unemployment rate.*

5%
10%
15%
20%
25%
30%
35%
40%
45%

Northumberland

Cumbria   Durham

**North
Yorkshire**

Lancashire   Leeds   Humberside

Cheshire

Gwynedd **Clwyd**   Derbyshire Nottinghamshire Lincolnshire

Staffordshire

Shropshire   Leicestershire   **Norfolk**

Powys   Warwickshire Northamptonshire   Cambridgeshire **Suffolk**

Dyfed   Hereford
and
Worcester

Gwent   Gloucestershire   Oxfordshire Buckinghamshire Hertfordshire   **Essex**

West
Glamorgan   Mid
Glamorgan

Avon   Berkshire   Middlesex

Wiltshire   **Surrey**   **Kent**

Somerset   Hampshire   West
Sussex   East
Sussex

Devon   Dorset   Isle
of
Wight

Cornwall

*Figure 2.5   The 97 family practitioner areas were, usually, amalgamations of
local government districts into which doctor's practices were grouped. These are
the areas for which migration rates were obtainable for the years between the
censuses. This is because National Health Service patients need to re-register
with a new doctor when they move.*

Box 2.2   Storing the geometry

The entire ward geometry of Britain was generalised using an algorithm that recorded a pair of vertices every five kilometres along the (original) length of each boundary. The list of boundaries was then stored with the following coded information: left ward, right ward, original length, number of vertices and (X, Y) vertex coordinates.

In spite of programming at the binary level the file was still over one megabyte in length.

A lookup table was also constructed as a binary file, giving the different area codes for each ward. These were its level II European region, county or metropolitan area or Scottish region, family practitioner committee area, postcode, local education authority, functional city area, local labour market area, travel-to-work area, local authority government district, parliamentary constituency and amalgamated office area.

A drawfile for a particular geography was produced using the draw library and a linked list of relevant boundaries constructed using the lookup table. The list for each area type was transformed into a series of polygons and the resulting group (path) tagged by its area identification.

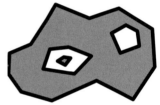

The software could handle over 10 000 units, and complex structures such as lakes within islands within lakes. The drawfile can be shaded or edited on the screen later, or further manipulated by other applications.

2.3   What are spaces?

Whether in prose or paint, on canvas or on digital photographs, on postcards or in poems, landscape representations are vehicles for the circulation of place through space and time. They take places out of their physical boundaries and move them around, shaping geographical imaginations.

(della Dora, 2011, p. 7)

Spaces are constructed from the relationships between places. Just as the individual attributes of people are not the main interest, so the collective attributes of single places do not hold the key to our understanding of society. Places are not things that can be rigidly defined and have a meaning of their own. They are abstract collections of people whose depiction can shed some light on the spatial social structure of our lives.

Aspatial views of society can capture many things. Gender relationships within households may not express a national geography. Those whose schooling was entirely in a private institution will never have experienced the spatial inequality of the state education sector, in which the place where you live can determine how well you learn (Figure 2.6).

In two dimensions we can see the environment in which people live as a whole, not only how they vote and work, but how everybody around them votes and works, and how other things about them are distributed – where they and their neighbours came from, how they live, who they are. Perhaps it is because this spatial structure is so strong, so well known, that we so often seek to find more ethereal aspatial relationships. We should first take a look at the wood before trying to classify the trees (as if they were not part of it).

There are, however, some fundamental problems to be overcome in trying to see the structure of society through its spatial apparitions.[6] To begin with, there is the problem of drawing a line around a group of people to be called a society, for which Britain was chosen for this book. Then, there is the question of how to cut up that space and what effects such divisions can have. Numerous lines have been drawn across Britain defining communities and cities, regions and villages (Figures 2.7, 2.8 and 2.9). How the spaces of interest can be rebuilt from the dissection of the nation is the central problem in relating places to people.

2.4   Drawing lines

> ... the 'whole' should be greater than the sum of the parts, or else
> we are dealing merely with an aggregate. This property of 'wholeness'
> of an object is reflected in its relative self-containment of activity.
> Boundaries around the object are characterised as zones of greater
> impermeability between the object and the outside world ...
>
> (Coombes et al., 1978, p. 1181)

Just as the line that surrounds Britain is dictated to us, so too are most of those that divide its people. Most seriously, for many purposes these lines divide the country into areas that contain far too many people to come anywhere near to the idea of the neighbourhood outlined above. Very small areas can also be too small to be a neighbourhood. At a ridiculously aggregated scale Britain is often divided into ten or eleven regions and numerous statistics periodically

---

[6] 'At least as long ago as 1950, it was demonstrated that not only the strengths of relationships between different variables change at different levels of aggregation but that, in some cases, the sign of the relationship also changes! Even worse, the effects of country-wide variations in size of areal units within the same level of the administrative hierarchy certainly provide variable levels of resolution and may also induce severe biases' (Rhind, 1975b, p. 9).

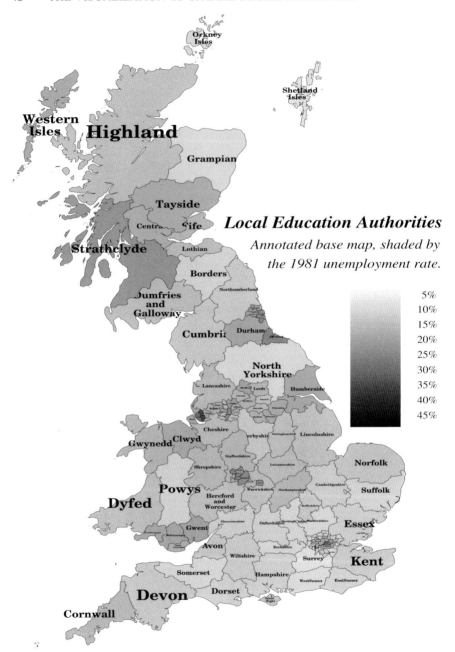

Figure 2.6   The 116 local education authorities were made up of metropolitan districts and counties. They were responsible for a large proportion of government expenditure. They are not used elsewhere in this book as few statistics were specifically provided for them. However, see how, on these boundaries, Liverpool appears as the area of highest unemployment in the country.

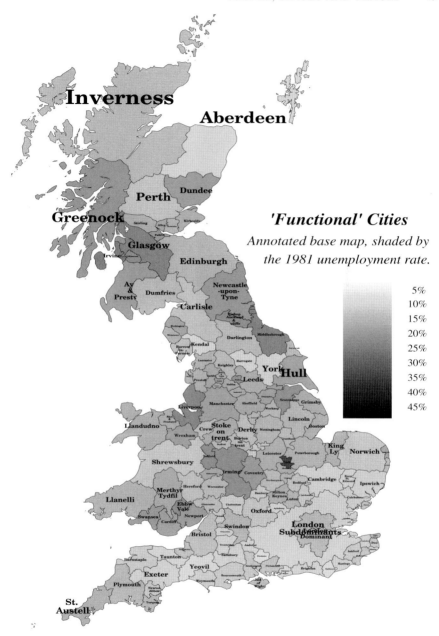

Figure 2.7   These are amalgamations of local labour market areas drawn around the 136 largest cities of the land, to include their spheres of influence. Notice how suddenly Corby appears as an unemployment blackspot for the first time in this series of maps, illustrating the effect of borderline decisions. Note how the South-Western Isles of Scotland change to 'Greenock'.

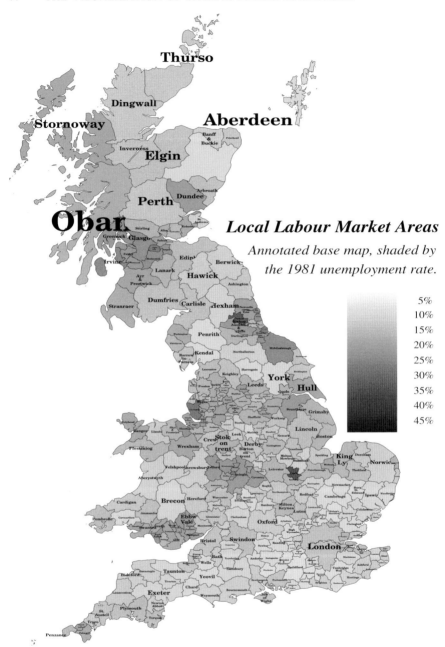

Local Labour Market Areas

Annotated base map, shaded by the 1981 unemployment rate.

5%
10%
15%
20%
25%
30%
35%
40%
45%

Figure 2.8   These 280 units were used, prior to 1991, by several previous studies of the census and other social statistics, and were specially created for that purpose. Gateshead has now joined Corby as two areas with equally high rates of joblessness on these boundaries, but other clusters begin to appear. The Southern Scottish islands are grouped under the label 'Oban' here.

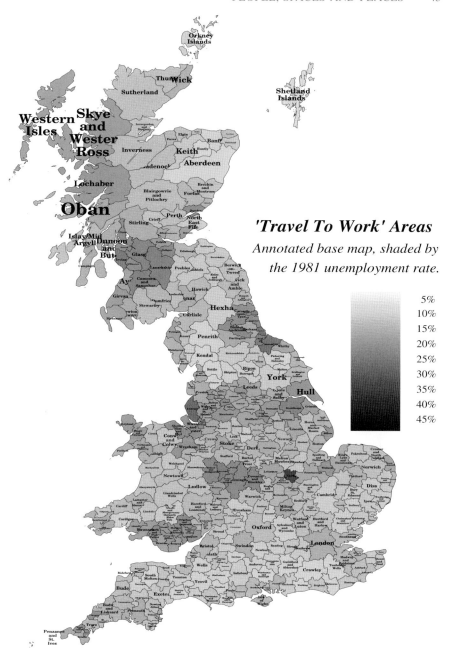

'Travel To Work' Areas

Annotated base map, shaded by the 1981 unemployment rate.

	5%
	10%
	15%
	20%
	25%
	30%
	35%
	40%
	45%

Figure 2.9   These 322 areas were created during the 1980s by Academics in Newcastle for the Department of Employment as being the most appropriate for the presentation of unemployment levels and other statistics. Notice that Corby is back as the single highest rate of unemployment and that the Scottish islands are now more accurately labelled with names like 'Islay'.

published for average people in these areas – for example the average Yorkshire and Humberside person.[7]

Other large areas used in the illustrations shown here often include the seventeen 1980s planning regions, the sixty-four counties, the four hundred and fifty-nine districts (Figure 2.10), the six hundred and thirty-three mainland parliamentary constituencies (Figure 2.11) and so on (Box 2.3). These areas define regions the size of small towns, often containing as many as a hundred thousand people, or more, or as few as a few dozen thousand. The spatial social structure hardly manifests itself here, for, in these areas, to sustain such areas, are people from all walks of life. Towns and cities are made up of the rich and poor. There are not rich and poor towns, at least not in the sense that there are rich and poor people.

What is more, the boundaries of the many areas for which we have data rarely coincide, neither with each other in space nor with themselves at different points in time. We are left with most of our geographical knowledge being mixed into a plethora of ephemeral places, the shapes and sizes of which do as much to alter the appearances of what is happening as do the numbers that are gathered about the people within them, even though they may be very accurate measures. The average for a large area destroys knowledge of the variations and defines people who do not exist – a bit of everyone and all of no one.

One way to try to overcome this confusion is to draw another set of lines, but this time defined not for the convenience of collection or to administer government or to distribute services, but on some functional grounds. These are called 'functional regions'. They are zones that try to encompass the areas within which people live and work or between which they tend to migrate. Although such exercises might usefully tell us something about the patterns of commuting and migration, they do not necessarily serve well for seeing social structure any better than the existing divisions – for several reasons.

One problem with the use of functional regions, in effect watersheds for people, is that they have tended to create even fewer places than have commonly been used and so the gross aggregation of communities is sustained, again creating the illusion that there is little spatial division. Administratively defined areas created for voting purposes tend at least to collect the same numbers of people within their boundaries; those made to operate local government at least divide London into boroughs. In contrast, these functionally defined entities exhibit some of the greatest variations in population of any set of boundaries in use.[8]

The very nature of 'functional' areas, created from flows of people, dispenses with the spatial aspect of society by creating a set of (semi-autonomous) places, which can be studied on their own, ranked and (supposedly) profitably listed

---

[7] Now called the average 'Yorkshire and the Humber' person as the boundaries have been tweaked a little. North East Lincolnshire is today in with the Humber (Dorling and Thomas, 2011, p. xx).

[8] One result is that when 'functional regions' were created by the centre I worked in when the first draft of the dissertation that became this book was written (CURDS), they were criticised for the statistical effects of widely varying denominator populations: 'The CURDS use of relative indices biases their account by emphasizing the growth of rural areas' (Savage, 1989, p. 255).

Box 2.3   The areal hierarchy

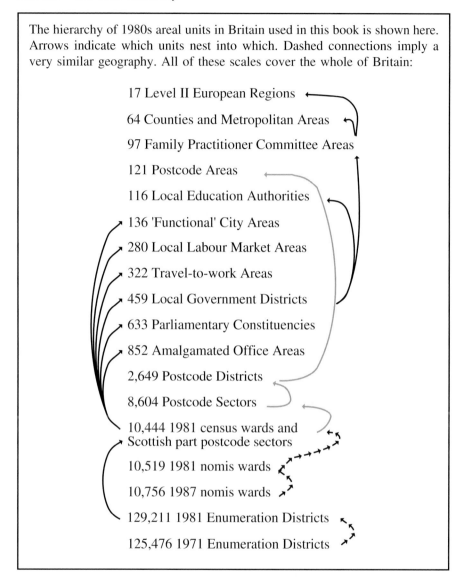

The hierarchy of 1980s areal units in Britain used in this book is shown here. Arrows indicate which units nest into which. Dashed connections imply a very similar geography. All of these scales cover the whole of Britain:

17 Level II European Regions

64 Counties and Metropolitan Areas

97 Family Practitioner Committee Areas

121 Postcode Areas

116 Local Education Authorities

136 'Functional' City Areas

280 Local Labour Market Areas

322 Travel-to-work Areas

459 Local Government Districts

633 Parliamentary Constituencies

852 Amalgamated Office Areas

2,649 Postcode Districts

8,604 Postcode Sectors

10,444 1981 census wards and Scottish part postcode sectors

10,519 1981 nomis wards

10,756 1987 nomis wards

129,211 1981 Enumeration Districts

125,476 1971 Enumeration Districts

in tables.[9] Functional regions are, in this sense at least, an aspatial concept, an attempt to take out the effect of movement and the relationship between places where the real differences are to be found.

---

[9] This I and colleagues have done several times in recent years to help illustrate the growth of the North–South divide, but which does not help illustrate that within each functionally defined area, within each city, town and possibly even within many larger villages, divides also have tended to grow.

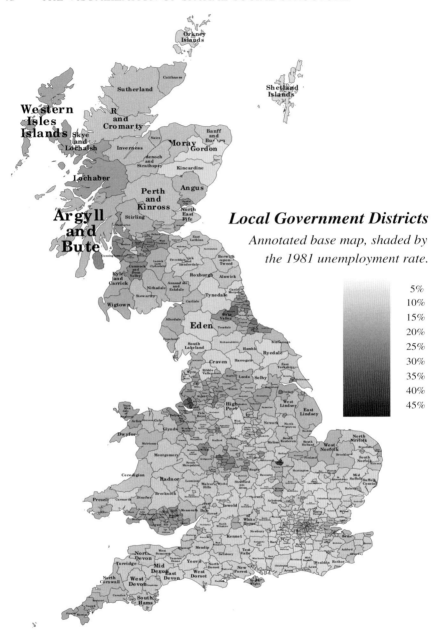

Figure 2.10   These are the 459 areas that local councils (elected by ward) governed. Spending was allocated by them and local taxes (the rates) collected. As a result, these are also the units for which most accurate mid-year population estimates were made. Many separate areas of high unemployment appear and the Southern Scottish islands are now under the name of 'Argyll and Bute'.

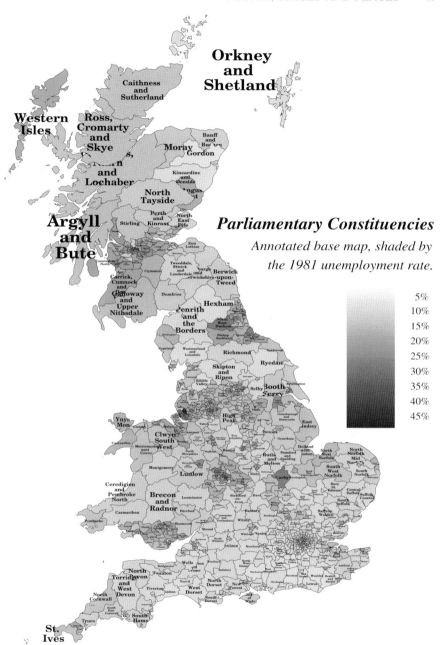

Figure 2.11   These 633 areas have the most equal levels of population of any set in Britain. Each has a single elected member of parliament. They are rarely used in mapping as no administrative role is given to them. Their apparently varying importance makes this type of map, when drawn on an equal land area basis, obviously inappropriate for depicting election results.

There has been a great deal of drawing of lines. It is worth seeing what is revealed when we begin rubbing them out, to see the real divisions, not the ones imposed by the cartographer's pen, administrator's necessity or geographer's algorithm. It is possible that just a few dozen carefully chosen regions do illustrate the diversity of society, areas each containing a similar population of around six hundred thousand people, for instance. However, to know if these can suffice for some purposes we have to first know what we are missing; we have to first include all the detail.

2.5   Picturing points

> *Place, therefore, refers to discrete if 'elastic' areas in which people can identify. The 'paths' and 'projects' of everyday life, to use the language of time-geography, provide the practical 'glue' for place.*
>
> *(Agnew, 1987, p. 28)*

To show the lived spatial distribution of people we need information about places that are at least as small as the communities we envisage containing the majority of everyday travels. These would be the areas on the map within which children travel to primary school, people shop and within which almost all the elderly live, the areas (increasingly) out of which younger adults commute to work or college.

If you think of the number of neighbourhoods in a town of two hundred thousand people, you should be able to count many distinct estates. Places ranging in size from as few as one hundred to as many as twenty thousand people are what many call neighbourhoods. Maybe it is with areas such as these that we need to begin with in order to build up a more comprehensive picture of the national spatial structure.

Even the smallest, conventionally used, administrative areas are usually too large for our purposes (Figure 2.12). The boundaries of smaller places may well be arbitrarily defined. However, that need not be a great problem if we are careful about our imaging techniques.[10] It is the relationship between the places in which we are interested. We should look at what places show collectively, not their individual characteristics – there are usually far too many of these to examine each one.

With enough small places we can create methods that produce results that are not artefacts of the arbitrary lines drawn on the map, devising techniques by which we can paint realistic pictures of social spaces. What are needed are techniques where it does not matter precisely how the points representing communities are defined. Using these techniques new maps of a slightly different set of areas

---

[10] 'It is not the areal units which are to blame. The difficulty is that the method of analysis used was inappropriate. This tautology is immediate. If the procedure used gives results which depend on the areal units used then, *ipso facto*, the procedure must be incorrect, and should be rejected *a priori*' (Tobler, 1989a, p. 115).

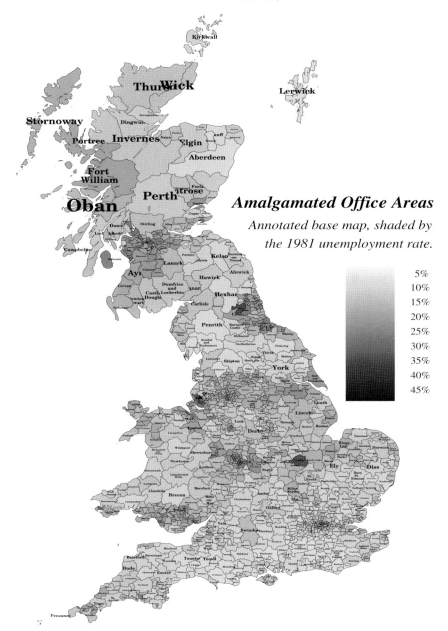

Amalgamated Office Areas

*Annotated base map, shaded by
the 1981 unemployment rate.*

5%
10%
15%
20%
25%
30%
35%
40%
45%

*Figure 2.12   These are 852 groups of unemployment office areas that super-
vised the allocation of benefit payments in the 1980s (former poor-law union
areas). They were combined to create a consistent base for comparison over
time. Notice the new shape to the area over which the eye-catching blackspot
of Corby is shaded. The Southern Scottish islands are now given names such as
'Campbeltown'.*

should look similar to each other despite the arbitrary local boundaries that differ between them.

For detailed social mapping to work, the most important factor about the localities that we use is that they vary little in the number of people that they contain. This is because larger areas tend to give a false sense of more uniformity within them. Again we require equal representation of people as a means to reduce the arbitrariness of the visualization. If you try to make sure that you give all people equal weight and don't draw graphics that place too much emphasis on a few, then you are less likely to become distracted by patterns that are arbitrary.

During the 1980s, at the same time as 'functional' regions were being created, other researchers were turning their hand to placing lattices over the land area of the British Isles and counting attributes of the number of people in each small square. This method of division was called grid mapping, and one kilometre squares were most often used.

Grid mapping is no less arbitrary a practice than any other technique, although the practicalities of its execution are the simplest. That it creates a stable set of units over time is a trivial defence of the method; any set of lines you draw and leave on a map is 'stable'. This type of spatial division can be seen as flawed for the visualization of human geography, partly because of the huge inequalities in populations located within the small unfamiliar areas it creates.[11]

Today a very fine population grid can be used and there is enough spare computing capacity not to worry that almost all the grid squares contain no people. Two decades ago computers were not so powerful, which was useful because it made you think more carefully about what you really wanted to do and see.

What if you wanted to see roughly ten thousand to one hundred thousand localities across all of Britain? Fortunately, the lowest tier of administrative geography gave and still gives us the first set – wards – and that of census geography the second – enumeration districts. There are other practical alternatives, for example postcode sectors, on which much market research is based, number over eight thousand (Figures 2.13, 2.14 and 2.15). However, in general the absolute wealth of information that is available at the time of the census for enumeration districts[12] and for wards between census years makes these places the natural choices from which to start painting spaces.

2.6 Population space

We do '.... high-pass filtering in our "mind's eye".'

(Tobler, 1989b, p. 19)

---

[11] 'More than this unfamiliarity though, is the deep-seated, if irrational, feeling that they are distinctly "unnatural", while "anything which wiggles" is "natural"' (Rhind, 1975a, p. 3).

[12] In more recent years census data have been disseminated by 'output areas', which differ a little in shape and size from the areas used to enumerate, but can be considered as largely equivalent here. They differ in being optimised to be more homogeneous, following the advice of Professor David Martin.

Figure 2.13   These 121 areas are based around postal towns and are identified by the first two letters of their postcode. They are coloured at random here to show what such random distributions look like. However, can you see any patterns? On the computer screen the map appears like a stained glass window, which shows how the printing process can change a picture.

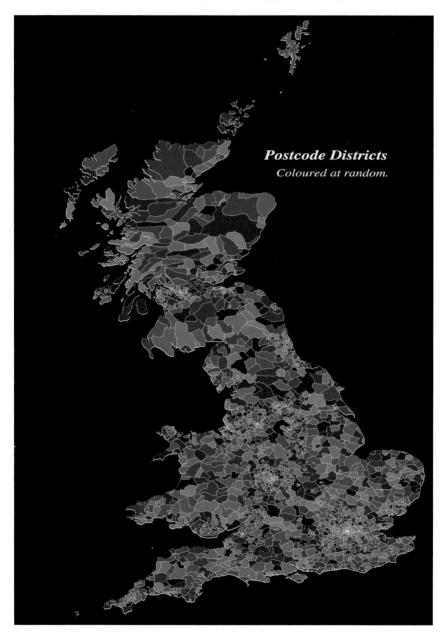

Figure 2.14   These 2649 postal district areas are drawn on an equal land area map. The first half of the full postcode gives the geographical resolution shown here. These areas are particularly useful in market research. They allow an intermediate level of geography, between wards and the previous higher amalgamations. The colouring is random to show what random is like.

Figure 2.15   These 8604 areas are drawn on an equal land area map. This is the finest level at which it is possible to map the boundaries of postcode unit geography. The first half and next digit of the postcode locate these areas. The image reveals a picture of population density as the white boundaries coalesce. Density is apparent despite the random colouring within all the white lines.

How should these pictures of spaces made up of people in places show us social structure? What should they look like and how should they reflect society? To answer these questions we must have some idea of what it is about the space we want to see – the nature of society.

People in space create a near continuum, especially if we view the distances between them in a relative sense. The spatial nature of our society is such that nearby places usually exhibit very similar characteristics in their populations, but occasionally they diverge widely. To see these patterns we have to stretch and squeeze the space of physical geography into becoming the landscape of human experience, opening up the cities and exorcising the empty space from the image.

The spatial nature of the society we live in holds more than divisions and continuity, trends and correlations. It is intricately patterned. Social patterns of power, control, deprivation and monotony are all reflected in spatial mosaics: rings of the wealthy, holes of the poor, lines of accessibility, enclaves of distinction. However, none of this would be seen if we did not seek to see it. We must know something of what we are looking for before we can know how to look.

Pictures of spatial social structure should have the power to reflect the complex tapestry and delicate lacework of the relations between people through the places in which they live and the spaces they create. All social organization must take place somewhere, and aspects of that somewhere strongly shape what structures are formed.[13]

At the time of first writing these words, a new software package called a geographic information system was just becoming more widely available (there are now several such software packages). It was very limited in what it could do and remains dedicated largely to replicating what it was possible to produce by hand in the past, just very quickly. However, because of these developments it was suggested back in 1984 that: 'Anyone at this stage writing their own software of this type is therefore foolhardy.'[14]

2.7   Adding time

A map of the earth with hundred-thousand-man-lost circles centred on England in 1850 establishes the most distant place, not as New Zealand, but somewhere around Moscow. The paths of least deaths at right angles to the circles draw another set of real longitudes and latitudes!

(Bunge, 1973, p. 286)

---

[13] 'Therefore, and this point is played down by Giddens, place is not just locale, as setting for activity and social interaction, but also location. The reproduction and transformation of social relations must take place somewhere' (Agnew, 1987, p. 27).

[14] The quote, tellingly, continued: 'All of these maps, however (with the exception of those by OPCS), deal with only up to about 1,000 or so zones on any one sheet of paper' (Rhind, Mounsey and Shepherd, 1984, p. 65).

Space and place is not enough. Just as the constraints of space are as important as those of time, without reintroducing time we only see a very partial picture, a snapshot. The detailed spatial structure to social life takes time to form and is deformed in time.[15]

Neighbourhoods are areas that have a spatial identity, just as a *generation* has a temporal identity. Without an identity the *cohort* in time is like the *locality* in space. People who live at the same time will more often live comparable lives of comparable quality, far more so for people who live at the same time and in the same places. These three-dimensional spacetime pockets of existence are the level of containment with which this book ends. They can be delimited by neighbourhoods and generations, bounded by a few hundred people and a few dozen years.

A history and geography of modern Britain, one that tries to tell the story of its entire people, would need as its building blocks sources that include most people. It is useful to base such a story around the concept of slowly changing, but ever changing, communities,[16] the definition of a community being a group of people with whom you share both space (locality) and time (cohort).

Localities are influenced by (and influence) other localities, just as cohorts are influenced by, and in turn influence, other cohorts. Cohorts are arranged in generations and locality in neighbourhoods, and together their intersection is a community. Every single person's community (locality and cohort combined) is unique to them, never quite perfectly overlapping anyone else's, most completely overlapping those with whom they share the most space and time. Every person's precise generation is unique to them, but is most similar to those with whom they share the most time, the same precise address and rough locality. The community spacetime bubbles within which we all fit all influence one another – but some far more than others.[17]

The definition of social structures in time is much less clear-cut than the definition in space. Communities merge into one another, evolve and – over the long run – disappear. The idea of rigid boundaries is even more ridiculous when applied to time than when applied to space. We use hours, days, weeks and years to regulate our lives and can see these as the scale on one of the axes of a three-dimensional spacetime continuum, a block of volume within which we

---

[15] 'Place is anchored in and takes its force from its physicality, and yet it often stretches far beyond it, it continues to travel, challenging space and time' (della Dora, 2011, p. 242).

[16] 'Cities do not only exist in the physical property of the cities, they also exist where the citizens of the city move. We are so property bound that we think a man lives at the address of his bed, even if he never visits it. The cities could be defined as their people as opposed to the property of the city. Using such a definition, the city has moved if the people evacuate it, say on a weekend holiday' (Bunge, 1973, p. 293).

[17] '. . . the "factors" causing political behaviour cannot just be added up in linear fashion (census class, census age, census ethnicity, etc.) to constitute an adequate explanation. To the contrary, it is how these factors "come together", take on meaning for people, and determine political outcomes that constitutes a satisfactory political analysis. In other words it is in places that causes produce the reasons that produce political behaviour' (Agnew, 1987, p. 213).

can reflect the temporal spatial structure of society and within which each of our lives is a path.

This book ends by suggesting that it is possible to produce a fixed three-dimensional image of an evolving, finely organised society. This would be a volume relating the times through which people live their lives to the places in which they live them. For now, creating the two-dimensional spaces to reflect simple two-dimensional instances of a complex structure is enough of a challenge, and the focus of the next chapter, but it is worth having an idea of where we are trying to get to if you want to appreciate why each step is being taken.

3

Artificial reality

*People ask me, 'What's so good about artificial reality?' and I say,
'What's so good about reality?'*
*Myron Krueger, quoted in New York, August 6, 1990 (Haggerty,
1991)*

3.1 Imagining reality

A model or picture of something is said to be impressive if it appears 'real', if it
looks like the original is imagined to be. Artificial implies not natural. Artificial
reality is a manufactured version of the original, created through imaginative
skill to show more about reality than is directly visible.

If we were to paint things just as we saw them, then our purpose would be
merely to store their likeness. Instead we wish to investigate their being. To bring
out more than the mere surface details of reality we must create images, artefacts,
which might not look directly like we initially imagine reality, but which tell us
far more about it and what lies beneath it.

Chemical models of molecules are a pioneering example. Not only do
they represent great magnifications of reality to a size we can see but, more
importantly, they distort, simplify and elaborate the object, to enhance our
understanding. The atoms making up the molecule are drawn as planets, brought
close together and linked with rods to imply connection. Unnecessary detail is
omitted and different elements are colour coded to aid interpretation. Highly
complex molecular modelling was part of the forefront of 1980s visualization,
high up then on the scientific computing agenda.

Transforming reality is an ancient occupation. From charting the Heavens to
depicting anatomy, we represent things not as they are but as we think they should

The Visualization of Spatial Social Structure, First Edition. Daniel Dorling.
© 2012 John Wiley & Sons, Ltd. Published 2012 by John Wiley & Sons, Ltd.

be understood. We create artificial realities partly through necessity – reality being too large, small or chaotic – but mainly to expedite understanding. In visualization we enhance reality.[1] The most important decision in so doing concerns what view of space to adopt as the basis for our pictures (Figure 3.1).

As the web expands and visual techniques develop, the contents of our imaginations are made visible to feed back into our minds and those of others in great loops of creativity. The ability to share our thoughts so vividly will tax our abilities to accept and cope with each other's artificial realities. Through seeing how others see, we all change our views.[2]

3.2 Abstract spaces

As a rule, the novel, dramatic character of cartograms may deceive unwary map readers. Great care and skill must be exercised when dealing with this particular type of map. The advantages of cartograms are substantial enough, however, that geographers would do well to gain sufficient sophistication to handle these maps effectively.

(Muehrcke, 1981, p. 27)

Viewed from a few hundred metres above the surface of the earth people appear like ants, milling around aimlessly. Another few hundred metres and we cannot see the people, only the buildings they have constructed, the land they had cleared and the roads and bridges that connect and separate these things (Figure 3.2). A few kilometres above the earth and all this evidence of human activity disappears; viewed from space orbit with the human eye we can just make out the pyramids of Egypt but not the Great Wall of China. We can see the blurs of a mega-city but no detail and the fans of excessive soil erosion out of great deltas, but not their upstream sources.

The creation of artificial spatial realities is necessary to our sense of self importance, to not seeing ourselves as akin to grains of sand on a beach. We often think that what we do is central to this world, that the thin slices of concrete we have placed upon its soil is evidence of great achievement and that our impact is of crucial importance to its future.

We may well be poisoning the air and oceans, but other than the smoke plume from a zoomed-in-on industrial complex or an algae bloom downstream of an

---

[1] 'The map is not some inferior but more convenient substitute for a globe. Map projections are not simply choices of lesser evils among distorting possibilities. On the contrary, the map allows the geographer to twist space into the condition he wishes. For purposes of finding lines of constant compass direction, the Mercator projection is far superior to the actual surface of the earth. The earth itself lacks the spatial property of having such lines being straight lines' (Bunge, 1966, p. 238).

[2] 'We found that the most effective maps may not be the most realistic, but are those which actually "distort" reality by eliminating information and by visually clarifying the topological and functional connections among geographical entities (e.g., a pocket subway map)' (Mills, 1981, p. 115).

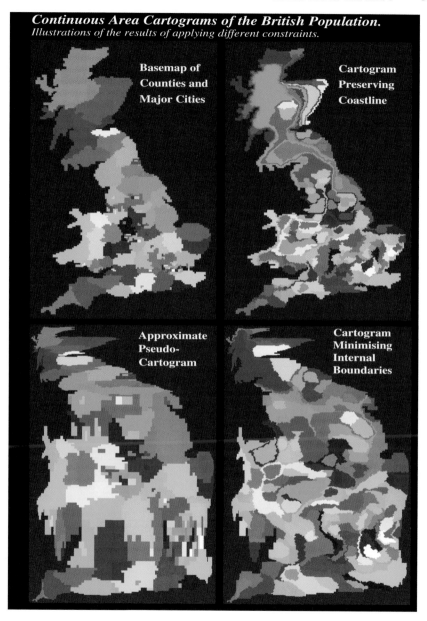

Figure 3.1   Top left is the basemap and next to that is a continuous area population cartogram preserving the physical coastline. Notice how the mass of northern English conurbations are squeezed, like toothpaste, up through the neck of Scotland. The bottom left is Tobler's pseudo-cartogram, where the marginal distributions, in this case of latitude and longitude, are equalised.

Figure 3.2   The bridges and tunnels shown here were from a list compiled by hand, based on 1984 road atlases; those and the 64 counties are shown on an equal land area map. Most major rivers and estuaries are spanned. These links are as important as shared boundaries to the geographical contiguity of the country. They are included in the matrices used to create most of the cartograms shown here.

over-farmed river, we can rarely see immediate physical evidence of our impact directly. Seeing that there are no longer trees requires knowing that there once were forests.

When we map our world we are mapping it for ourselves: to navigate it, control it and understand it as it relates to us.[3] The objective view of the earth, the blue-green blob seen from the moon, tells us nothing of the world of people, who might as well not be there for all we can see. To see ourselves on this planet we had to create first the abstract spaces upon which our paths could be drawn (Figure 3.3).

Early maps of the world were centred on the religious capitals, the land was magnified where most people were known to live and most detail could be drawn.[4] There were maps of kingdoms and empires, spaces that contained land – the land that contained people. The world was flat, as you could not sail around it (at least not around the back). Rivers were drawn as wide barriers, mountains enlarged as impassable obstacles. Today roads are drawn far wider than their real width.

Map projections were first deliberately devised to aid navigation; straight lines on the map maintained their compass orientation (Box 3.1). The shape of the world changed again, and suddenly it was full of oceans and seas where once the land had crowded out the space for water so as to fit in the names of places

Box 3.1   The Mercator projection

The Mercator Projection maintains all compass directions as straight lines and was, thus, extremely useful in an age of maritime navigation. It distorts areas considerably. Greenland is very much magnified while Africa is relatively shrunk. The shape of the areas is also altered, but this is always inevitable, to some extent, in flat representations of the surface of the Earth.

It is surprising that, centuries after its inception, this image (and other images like it) should still present the accepted view of the world, a series of lines, dividing land from water, which mediaeval explorers used to chart their way over the oceans to distant coasts.

---

[3] '... simple mechanical accuracy in maps is not enough. Map makers should provide psychologic, or call it aesthetic, accuracy as well .... [a conventional] map could not be much more deceiving than it is even if a conscious effort were made to make it so' (Williams, 1976, p. 216).

[4] 'In ancient times and the middle ages, maps were highly subjective. No impersonal codes and conventions. No uniform scale, orientation or even distances' (Hagen, 1982, p. 326).

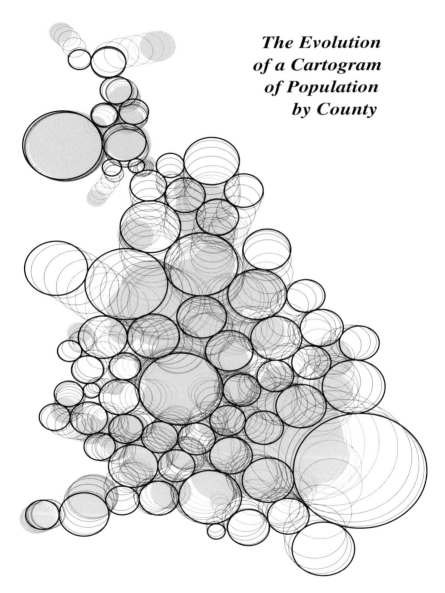

*The Evolution
of a Cartogram
of Population
by County*

*Figure 3.3   Ten stages in the development of a cartogram based on the 1981
census populations of 64 counties are shown, illustrating the mechanics of some of
the software. The picture gives an idea of the dynamics of the cartogram algorithm
and the interplay of each area pushing for space. The West Midlands circle in the
centre can be seen to have hardly moved at all.*

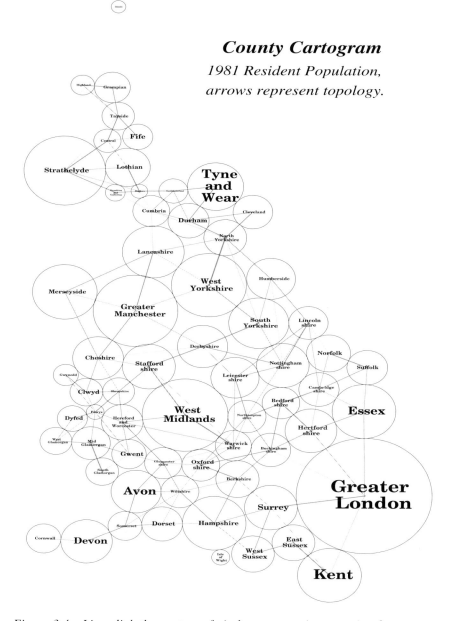

County Cartogram

1981 Resident Population,
arrows represent topology.

Figure 3.4   Lines link the centres of circles representing counties that are contiguous. The width of the lines is in proportion to the strength of continuity (which is governed by the lengths of common boundaries). Notice that the Isle of Wight and combined Scottish islands at the top are not connected to the mainland. The West Midlands has the least neighbours of any inland area – three.

and the courses of rivers.[5] The shape of the physical world had not changed but the world of man had, just as it has in the pictures drawn here (Figure 3.4). Trade and conquest by sea became paramount and the images changed to reflect this. How we saw our world changed as we changed our world and how we had changed it then changed us.

As all the lands were overrun by the former seafaring people, their area had to be subdivided and their actual size grew in importance. The shapes on the maps changed again as land area was maintained at the expense of compass direction.[6] Oceans, now easily traversed, shrunk as they were cut out of the atlas. The parcels of farm land to be settled, fought over and traded became more clearly depicted.[7] The centres of the maps moved from Mecca and Jerusalem to Venice and London as the centres of power moved, and the names were of dominions rather than provinces.[8]

The century became a time of air travel and world wars, of twenty-four-hour money markets and starvation on continental scales. All this changed how we viewed the world. Now the shape of the world is changing again, but to no single accepted projection. There are numerous one-world projections, from those that claim the peoples of the world are best represented by their land areas to those in which distance represents the shortest paths of ballistic missiles.

The shape of our world has always been an artificial reality of the times, whatever religious or scientific accuracy was claimed for it. We shape our world for our own purposes, to see where we are ourselves and to see what each other has. We are still imagining the abstract spaces to draw ourselves upon. They have never been naturally given.

3.3   Area cartograms

A cartogram is a combination map and graph.

(Wilkie, 1976, p. 1)

---

[5] 'Maps prepared using these [equal population] transformations, however, from many points of view are more realistic than the conventional maps used in geography' (Tobler, 1961, pp. 162–163).

[6] 'We need to recognize unequivocally that the map is a socially constituted image and our definition of the artefact itself should reflect that recognition' (Harley, 1990, p. 6).

[7] '... towns occupy spaces on the map – even allowing for cartographic convention – far in excess of their sizes on the ground. Castle signs, too, signifying feudal rank and military might, are sometimes larger than signs for villages, despite the lesser area they occupied on the ground' (Harley, 1988, pp. 292–294).

[8] A few pages on, however: 'Maps as an impersonal type of knowledge tend to "desocialise" the territory they represent. They foster the notion of a socially empty space. The abstract quality of the map, embodied as much in the lines of a fifteenth century Ptolemaic projection as in the contemporary images of computer cartography, lessens the burden of conscience about people in the landscape. Decisions about the exercise of power are removed from the realm of immediate face-to-face contacts' (Harley, 1988, p. 303).

Cartograms are maps in which the particular distortion chosen is made explicit. Area cartograms are drawn so that the areas representing places on the paper or screen are in proportion to a specific aspect of those places. The aspect most commonly chosen has been total human population; another age might have chosen only adult men.

Population cartograms give another shape to the world. By choosing to draw the surface of the earth as a population cartogram, we give everyone equal representation in the image (Figures 3.5, 3.6 and 3.7). In the process we lose much that is familiar, but then we do not learn through familiarity, and familiarity often deters introspection.

Population cartograms far outnumber any other kind. They were first drawn around a hundred and twenty years ago. In the 1960s algorithms were developed to construct them by machine, as their manual creation has always been immensely tedious. Most used in the 1980s were generated manually, often very artistically, although interesting mechanical means had also developed for their production. Much effort went into this pioneering work, because much was hoped of the media. What can be done today is largely a realization of what people were trying to achieve in the past, what they argued over and how those disputes were resolved.[9]

The depiction of electoral geography is a frequent use of population cartograms. Here the population base is often the electorate, although it can be political representatives when every area with one representative is given equal size. On any traditional map of an urbanised country, the majority of political constituencies are literally not visible to the naked eye (Figures 3.8 and 3.9). The problem is particularly acute in countries such as Canada and Australia, but still fundamental in all other regions of the world. The argument is not that the conventional map distorts the message; it is that it cannot correctly convey even a very small fraction of it.

Numerous insets, and insets within insets, or dynamic zooming (pushing apart two fingers on the touch-screen) could be employed to try and see what is going on, but they cannot form what is required – a single gestalt image, a unique impression (Figure 3.10). By 2010 The British Broadcasting Corporation had begun using cartograms in their election coverage.[10]

---

[9] 'The Tobler algorithm is regarded as imaginative but highly inaccurate, slow due to the number of iterations required by the algorithm, and guilty of producing an over generalized end product. ... This led Nicholas Chrisman to write a competing algorithm which uses a different distorted plane approach. In this scheme, each region or polygon has an amount of "force" applied to it based on the variable's value being mapped (Dougenik *et al.*, 1983). The implementation of the Chrisman algorithm ... currently exists as part of the mainframe GIS package ODYSSEY ... ' (Torguson, 1990, p. 20). Both of these algorithms have now been superseded by the Gastner–Newman algorithm. See footnote 12 below in this chapter.

[10] 'Television companies might argue that these examples of animated cartography are too complex for the average viewer to understand. But today's viewers are much more sophisticated than their parents who were happy to watch a cardboard pendulum be pushed across the screen. It is true that it will take some time to explain the graphics. But when have television viewers ever had more time, than when they are waiting for the slow trickle of results to come in on election night' (Dorling, 1994, p. 21).

Figure 3.5   The 459 districts are labelled for identification against an alphabet-ical index. The numbers are automatically scaled to the size of the land area, so that at least some can be read, rather than all being too small to see, and to illustrate the variation in size between the districts. This scaling highlights how it is the least densely populated districts that are most prominent.

Population cartograms will continue to gain in popularity as they become easier to employ and better understood in general.[11] For visualizing the spatial distributions of social structure there is no alternative, if we wish to see the detail of substance. A traditional map can take many projections, and so too can population cartograms. An infinite number of correct population cartograms can be constructed for any aspect of any set of places.

3.4   The nature of space

Vision is therefore, first and foremost, an information-processing task, but we cannot think of it just as a process. For if we are capable of knowing what is where in the world, our brains must somehow be capable of representing this information – in all its profusion of color and form, beauty, motion and detail.

(Marr, 1982, p. 3)

Cartograms are normally made by slowly stretching some parts while squashing others, until the places' sizes are in proportion to their populations, instead of being in proportion to their land area.

This process can be tempered by deciding that the topology of the space should be preserved throughout the transformation. In other words, these places, which were neighbouring, should remain so after transformation and those that were not so should not become neighbours. Most creators of cartograms aim to create a topologically and geometrically correct contiguous area cartogram. Even so, it is still possible to create a multitude of these for any given area. Recently Gastner and Newman's algorithm has allowed the widespread creation of cartograms, which also appear conformal (maintaining compass directions locally), reducing this multitude down, but it remains useful to look at other options and other constraints that may be helpful.[12]

Further constraints can be added to creating cartograms other than preserving topology, scaling areas appropriately and attempting to maintain compass directions where possible. The most common are that the outer boundary of the area be preserved and that the lengths of interior boundaries be minimised, so creating a cartogram, the shape of which looks familiar and whose interior is

---

[11] Not all writers favour the use of cartograms. Despite the format being very rare even in the 1930s there were objections to these types of graphic: 'Such maps, however, are not superior to the dot or bar maps ... for showing distributions of size. In many cases the method ... would result in the states being so distorted that little if any resemblance of their true shapes would remain, and even their relative positions would be inaccurate. It is much more difficult to compare the irregular areas on such maps than it is to compare either circles or bars' (Riggleman, 1936, pp. 179–180).

[12] Dorling, Barford and Newman, (2006). See also the on-line atlas www.worldmapper.org.

Local Authority Districts (1981)

1 Aberconwy
2 Aberdeen City
3 Adur
4 Afan
5 Allerdale
6 Alnwick
7 Alyn and Deeside
8 Amber Valley
9 Angus
10 Annandale and Eskdale
11 Arfon
12 Argyll and Bute
13 Arun
14 Ashfield
15 Ashford
16 Aylesbury Vale
17 Babergh
18 Badenoch and Strathspey
19 Banff and Buchan
20 Barking and Dagenham
21 Barnet
22 Barnsley
23 Barrow-in-Furness
24 Basildon
25 Basingstoke and Deane
26 Bassetlaw
27 Bath
28 Bearsden and Milngavie
29 Berwick-upon-Tweed
30 Berwickshire
31 Beverley
32 Bexley
33 Birmingham
34 Blaby
35 Blackburn
36 Blackpool
37 Blaenau Gwent
38 Blyth Valley
39 Bolsover
40 Bolton
41 Boothferry
42 Boston
43 Bournemouth
44 Bracknell
45 Bradford
46 Braintree
47 Breckland
48 Brecknock
49 Brent
50 Brentwood
51 Bridgnorth
52 Brighton
53 Bristol
54 Broadland
55 Bromley
56 Bromsgrove

57 Broxbourne
58 Broxtowe
59 Burnley
60 Bury
61 Caithness
62 Calderdale
63 Cambridge
64 Camden
65 Cannock Chase
66 Canterbury
67 Caradon
68 Cardiff
69 Carlisle
70 Carmarthen
71 Carrick
72 Castle Morpeth
73 Castle Point
74 Ceredigion
75 Charnwood
76 Chelmsford
77 Cheltenham
78 Cherwell
79 Chester
80 Chester-le-Street
81 Chesterfield
82 Chichester
83 Chiltern
84 Chorley
85 Christchurch
86 City of London
87 Clackmannan
88 Cleethorpes
89 Clydebank
90 Colchester
91 Colwyn
92 Congleton
93 Copeland
94 Corby
95 Cotswold
96 Coventry
97 Craven
98 Crawley
99 Crewe and Nantwich
100 Croydon
101 Cumbernauld and Kilsyth
102 Cunnock and Doon Valley
103 Cunninghame
104 Cynon Valley
105 Dacorum
106 Darlington
107 Dartford
108 Daventry
109 Delyn
110 Derby
111 Derwentside
112 Dinefwr
113 Doncaster
114 Dover
115 Dudley
116 Dumbarton
117 Dundee City
118 Dunfermline
119 Durham
120 Dwyfor
121 Ealing
122 Easington
123 East Cambridgeshire
124 East Devon
125 East Hampshire
126 East Hertfordshire
127 East Kilbride
128 East Lindsey

129 East Lothian
130 East Northamptonshire
131 East Staffordshire
132 East Yorkshire
133 Eastbourne
134 Eastleigh
135 Eastwood
136 Eden
137 Edinburgh City
138 Ellesmere Port and Neston
139 Elmbridge
140 Enfield
141 Epping Forest
142 Epsom and Ewell
143 Erewash
144 Ettrick and Lauderdale
145 Exeter
146 Falkirk
147 Fareham
148 Fenland
149 Forest Heath
150 Forest of Dean
151 Fylde
152 Gateshead
153 Gedling
154 Gillingham
155 Glanford
156 Glasgow City
157 Gloucester
158 Glyndwr
159 Gordon
160 Gosport
161 Gravesham
162 Great Grimsby
163 Great Yarmouth
164 Greenwich
165 Guildford
166 Hackney
167 Halton
168 Hambleton
169 Hamilton
170 Hammersmith and Fulham
171 Harborough
172 Haringey
173 Harlow
174 Harrogate
175 Harrow
176 Hart
177 Hartlepool
178 Hastings
179 Havant
180 Havering
181 Hereford
182 Hertsmere
183 High Peak
184 Hillingdon
185 Hinckley and Bosworth
186 Holderness
187 Horsham
188 Hounslow
189 Hove
190 Huntingdon
191 Hyndburn
192 Inverclyde
193 Inverness
194 Ipswich
195 Isles of Scilly
196 Islington
197 Islwyn
198 Kennet

199 Kensington and Chelsea
200 Kerrier
201 Kettering
202 Kilmarnock and Loudoun
203 Kincardine and Deeside
204 Kingston upon Hull
205 Kingston upon Thames
206 Kingswood
207 Kirkcaldy
208 Kirklees
209 Knowsley
210 Kyle and Carrick
211 Lambeth
212 Lanark(now Clydesdale)
213 Lancaster
214 Langbaurgh
215 Leeds
216 Leicester
217 Leominster
218 Lewes
219 Lewisham
220 Lichfield
221 Lincoln
222 Liverpool
223 Llanelli
224 Lliw Valley
225 Lochaber
226 Luton
227 Macclesfield
228 Maidstone
229 Maldon
230 Malvern Hills
231 Manchester
232 Mansfield
233 Medina
234 Meirionnydd
235 Melton
236 Mendip
237 Merthyr Tydfil
238 Merton
239 Mid Bedfordshire
240 Mid Devon
241 Mid Suffolk
242 Mid Sussex
243 Middlesbrough
244 Midlothian
245 Milton Keynes
246 Mole Valley
247 Monklands
248 Monmouth
249 Montgomery
250 Moray
251 Motherwell
252 Nairn
253 Neath
254 New Forest
255 Newark
256 Newbury
257 Newcastle upon Tyne
258 Newcastle-under-Lyme
259 Newham
260 Newport
261 Nithsdale
262 North Bedfordshire
263 North Cornwall
264 North Devon

265 North Dorset
266 North East Derbyshire
267 North East Fife
268 North Hertfordshire
269 North Kesteven
270 North Norfolk
271 North Shropshire
272 North Tyneside
273 North Warwickshire
274 North West Leicestershire
275 North Wiltshire
276 Northampton
277 Northavon
278 Norwich
279 Nottingham
280 Nuneaton and Bedworth
281 Oadby and Wigston
282 Ogwr
283 Oldham
284 Orkney Islands
285 Oswestry
286 Oxford
287 Pendle
288 Penwith
289 Perth and Kinross
290 Peterborough
291 Plymouth
292 Poole
293 Portsmouth
294 Preseli
295 Preston
296 Purbeck
297 Radnor
298 Reading
299 Redbridge
300 Redditch
301 Reigate and Banstead
302 Renfrew
303 Restormel
304 Rhondda
305 Rhuddlan
306 Rhymney Valley
307 Ribble Valley
308 Richmond upon Thames
309 Richmondshire
310 Rochdale
311 Rochester upon Medway
312 Rochford
313 Ross and Cromarty
314 Rossendale
315 Rother
316 Rotherham
317 Roxburgh
318 Rugby
319 Runnymede
320 Rushcliffe
321 Rushmoor
322 Rutland
323 Ryedale
324 Salford
325 Salisbury
326 Sandwell
327 Scarborough
328 Scunthorpe
329 Sedgefield
330 Sedgemoor

331 Sefton
332 Selby
333 Sevenoaks
334 Sheffield
335 Shepway
336 Shetland Islands
337 Shrewsbury and Atcham
338 Skye and Lochalsh
339 Slough
340 Solihull
341 South Bedfordshire
342 South Bucks
343 South Cambridgeshire
344 South Derbyshire
345 South Hams
346 South Herefordshire
347 South Holland
348 South Kesteven
349 South Lakeland
350 South Norfolk
351 South Northamptonshire
352 South Oxfordshire
353 South Pembrokeshire
354 South Ribble
355 South Shropshire
356 South Staffordshire
357 South Tyneside
358 South Wight
359 Southampton
360 Southend-on-Sea
361 Southwark
362 Spelthorne
363 St Albans
364 St Edmundsbury
365 St Helens
366 Stafford
367 Staffordshire Moorlands
368 Stevenage
369 Stewartry
370 Stirling
371 Stockport
372 Stockton-on-Tees
373 Stoke-on-Trent
374 Stratford-on-Avon
375 Strathkelvin
376 Stroud
377 Suffolk Coastal
378 Sunderland
379 Surrey Heath
380 Sutherland
381 Sutton
382 Swale
383 Swansea
384 Taff-Ely
385 Tameside
386 Tamworth
387 Tandridge
388 Taunton Deane
389 Teesdale
390 Teignbridge
391 Tendring
392 Test Valley
393 Tewkesbury
394 Thamesdown

395 Thanet
396 The Wrekin
397 Three Rivers
398 Thurrock
399 Tonbridge and Malling
400 Torbay
401 Torfaen
402 Torridge
403 Tower Hamlets
404 Trafford
405 Tunbridge Wells
406 Tweeddale
407 Tynedale
408 Uttlesford
409 Vale of Glamorgan
410 Vale of White Horse
411 Vale Royal
412 Wakefield
413 Walsall
414 Waltham Forest
415 Wandsworth
416 Wansbeck
417 Wansdyke
418 Warrington
419 Warwick
420 Watford
421 Waveney
422 Waverley
423 Wealden
424 Wear Valley
425 Wellingborough
426 Welwyn Hatfield
427 West Derbyshire
428 West Devon
429 West Dorset
430 West Lancashire
431 West Lindsey
432 West Lothian
433 West Norfolk
434 West Oxfordshire
435 West Somerset
436 West Wiltshire
437 Western Isles
438 Westminster,City of
439 Weymouth and Portland
440 Wigan
441 Wigtown
442 Wimborne
443 Winchester
444 Windsor and Maidenhead
445 Wirral
446 Woking
447 Wokingham
448 Wolverhampton
449 Woodspring
450 Worcester
451 Worthing
452 Wrexham Maelor
453 Wychavon
454 Wycombe
455 Wyre
456 Wyre Forest
457 Yeovil
458 Ynys Mon-Isle of Anglesey
459 York

Figure 3.6   The official names of all the 459 districts that existed in 1981 are shown, sorted and indexed alphabetically. Places can be identified and individual districts found on the maps and cartograms with these numbers shown on them by using this list. Note how if all the names of all the districts are to be included they can only just be printed when evenly distributed.

least convoluted. While it is possible to achieve both these aims simultaneously, possibly also producing a unique solution, they are somewhat contradictory aims.

Maintenance of the original perimeter dramatically restricts simplification of the internal boundaries. A population cartogram of Britain can be produced that precisely preserves the original coastline, but a confused internal structure results.

Local Authority Districts

(1981 Resident Population)

Arrows Represent Topology

Figure 3.7   The 459 districts are labelled for identification. Lines link the centres of circles connecting districts that are contiguous. The width of the lines is in proportion to the strength of the contiguity. The vast majority of previously contiguous districts are still touching each other, even in difficult cases such as the cluster around Glasgow city (number 156) or Birmingham (number 33).

Figure 3.8   The 633 Westminster parliamentary constituencies are labelled for identification against an alphabetical index. The numbers are automatically scaled to the land area, again so that at least some can be read rather than all being too small to see, and they emphasise the need for a more equitable projection. Notice the mass of unreadable numbers that merge together within London.

A cellular automaton algorithm was used to achieve this.[13] In this work the focus is instead on creating the simplest population cartogram, only roughly following the physical outline of Britain, so that the patterns are depicted with the least visual distortion and so that the greatest internal detail can be included through the many thousands of individual areas being shown (Figure 3.11).

The shapes of internal places in most cartograms in this book have been made circular (Box 3.2), and hence as simple to gauge in terms of their area/importance as possible. Strictly speaking, the contiguity and topological constraints are now broken, but in practice most places still bordered their former neighbours after transformation. Various methods could be employed to make the cartogram, once

Box 3.2   The algorithm at work

The algorithm that was developed to create the area cartograms worked by repeatedly applying a series of forces to the circles representing the places. Circles attract those they are topologically adjacent to, the strength of this attraction being greater the larger the distance is between them and the longer their common boundary.

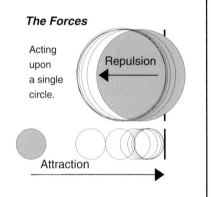

They repel those that they overlap, with a strength proportional to the overlap. Friction is applied to prevent unsatisfactory local solutions being settled too soon. The repulsion factor must always be slightly greater than the attraction or else (where, for example, each of four zones are all connected to the other three) an overlap will always remain.

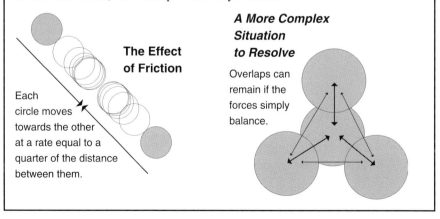

_____

[13] The algorithm is described in Dorling (1996).

Parliamentary Constituencies (1981)

1 Aberavon
2 Aberdeen North
3 Aberdeen South
4 Aldershot
5 Aldridge - Brownhills
6 Altrincham and Sale
7 Alyn and Deeside
8 Amber Valley
9 Angus East
10 Argyll and Bute
11 Arundel
12 Ashfield
13 Ashford
14 Ashton-under-Lyne
15 Aylesbury
16 Ayr
17 Banbury
18 Banff and Buchan
19 Barking
20 Barnsley Central
21 Barnsley East
22 Barnsley West and Penistone
23 Barrow and Furness
24 Basildon
25 Basingstoke
26 Bassetlaw
27 Bath
28 Batley and Spen
29 Battersea
30 Beaconsfield
31 Beckenham
32 Berwick-upon-Tweed
33 Bethnal Green and Stepney
34 Beverley
35 Bexhill and Battle
36 Bexley Heath
37 Billericay
38 Birkenhead
39 Birmingham Edgbaston
40 Birmingham Erdington
41 Birmingham Hall Green
42 Birmingham Hodge Hill
43 Birmingham Ladywood
44 Birmingham Northfield
45 Birmingham Perry Barr
46 Birmingham Selly Oak
47 Birmingham Small Heath
48 Birmingham Sparkbrook
49 Birmingham Yardley
50 Bishop Auckland
51 Blaby
52 Blackburn
53 Blackpool North
54 Blackpool South
55 Blaenau Gwent
56 Blaydon
57 Blyth Valley
58 Bolsover
59 Bolton North East
60 Bolton South East
61 Bolton West
62 Booth Ferry
63 Bootle
64 Bosworth
65 Bournemouth East
66 Bournemouth West
67 Bow and Poplar
68 Bradford North
69 Bradford South
70 Bradford West
71 Braintree
72 Brecon and Radnor
73 Brent East
74 Brent North
75 Brent South
76 Brentford and Isleworth
77 Brentwood and Ongar
78 Bridgend
79 Bridgwater
80 Bridlington
81 Brigg and Cleethorpes
82 Brighton Kemptown
83 Brighton Pavilion
84 Bristol East
85 Bristol North West
86 Bristol South
87 Bristol West
88 Bromsgrove
89 Broxbourne
90 Broxtowe
91 Buckingham
92 Burnley
93 Burton
94 Bury North
95 Bury South
96 Bury St. Edmunds
97 Caernarfon
98 Caerphilly
99 Caithness and Sutherland
100 Calder Valley
101 Cambridge
102 Cannock and Burntwood
103 Canterbury
104 Cardiff Central
105 Cardiff North
106 Cardiff South and Penarth
107 Cardiff West
108 Carlisle
109 Carmarthen
110 Carrick, Cumnock and Doon Valley
111 Carshalton and Wallington
112 Castle Point
113 Central Fife
114 Central Suffolk
115 Ceredigion and Pembroke North
116 Cheadle
117 Chelmsford
118 Chelsea
119 Cheltenham
120 Chertsey and Walton
121 Chesham and Amersham
122 Chesterfield
123 Chichester
124 Chingford
125 Chipping Barnet
126 Chislehurst
127 Chorley
128 Christchurch
129 Cirencester and Tewkesbury
130 City of Chester
131 City of Durham
132 Clackmannan
133 Clwyd North West
134 Clwyd South West
135 Clydebank and Milngavie
136 Clydesdale
137 Colne Valley
138 Congleton
139 Conwy
140 Copeland
141 Corby
142 Coventry North East
143 Coventry North West
144 Coventry South East
145 Coventry South West
146 Crawley
147 Crewe and Nantwich
148 Crosby
149 Croydon Central
150 Croydon North East
151 Croydon North West
152 Croydon South
153 Cumbernauld and Kilsyth
154 Cunninghame North
155 Cunninghame South
156 Cynon Valley
157 Dagenham
158 Darlington
159 Dartford
160 Daventry
161 Davyhulme
162 Delyn
163 Denton and Reddish
164 Derby North
165 Derby South
166 Devizes
167 Dewsbury
168 Don Valley
169 Doncaster Central
170 Doncaster North
171 Dover
172 Dudley East
173 Dudley West
174 Dulwich
175 Dumbarton
176 Dumfries
177 Dundee East
178 Dundee West
179 Dunfermline East
180 Dunfermline West
181 Ealing Acton
182 Ealing North
183 Ealing Southall
184 Easington
185 East Berkshire
186 East Hampshire
187 East Kilbride
188 East Lindsey
189 East Lothian
190 East Surrey
191 Eastbourne
192 Eastleigh
193 Eastwood
194 Eccles
195 Eddisbury
196 Edinburgh Central
197 Edinburgh East
198 Edinburgh Leith
199 Edinburgh Pentlands
200 Edinburgh South
201 Edinburgh West
202 Edmonton
203 Ellesmere Port and Neston
204 Elmet
205 Eltham
206 Enfield North
207 Enfield Southgate
208 Epping Forest
209 Epsom and Ewell
210 Erewash
211 Erith and Crayford
212 Esher
213 Exeter
214 Falkirk East
215 Falkirk West
216 Falmouth and Camborne
217 Fareham
218 Faversham
219 Feltham and Heston
220 Finchley
221 Folkestone and Hythe
222 Fulham
223 Fylde
224 Gainsborough and Horncastle
225 Galloway and Upper Nithsdale
226 Gateshead East
227 Gedling
228 Gillingham
229 Glanford and Scunthorpe
230 Glasgow Cathcart
231 Glasgow Central
232 Glasgow Garscadden
233 Glasgow Govan
234 Glasgow Hillhead
235 Glasgow Maryhill
236 Glasgow Pollock
237 Glasgow Provan
238 Glasgow Rutherglen
239 Glasgow Shettleston
240 Glasgow Springburn
241 Gloucester
242 Gordon
243 Gosport
244 Gower
245 Grantham
246 Gravesham
247 Great Grimsby
248 Great Yarmouth
249 Greenock and Port Glasgow
250 Greenwich
251 Guildford
252 Hackney North and Stoke Newington
253 Hackney South and Shoreditch
254 Halesowen and Stourbridge
255 Halifax
256 Halton
257 Hamilton
258 Hammersmith
259 Hampstead and Highgate
260 Harborough
261 Harlow
262 Harrogate
263 Harrow East
264 Harrow West
265 Hartlepool
266 Harwich
267 Hastings and Rye
268 Havant
269 Hayes and Harlington
270 Hazel Grove
271 Hemsworth
272 Hendon North
273 Hendon South
274 Henley
275 Hereford
276 Hertford and Stortford
277 Hertsmere
278 Hexham
279 Heywood and Middleton
280 High Peak
281 Holborn and St. Pancras
282 Holland with Boston
283 Honiton
284 Hornchurch
285 Hornsey and Wood Green
286 Horsham
287 Houghton and Washington
288 Hove
289 Huddersfield
290 Huntingdon
291 Hyndburn
292 Ilford North
293 Ilford South
294 Inverness, Nairn and Lochaber
295 Ipswich
296 Isle of Wight
297 Islington North
298 Islington South and Finsbury
299 Islwyn
300 Jarrow
301 Keighley
302 Kensington
303 Kettering
304 Kilmarnock and Loudoun
305 Kincardine and Deeside
306 Kingston-upon-Hull East
307 Kingston-upon-Hull North
308 Kingston-upon-Hull West
309 Kingston-upon-Thames
310 Kingswood
311 Kirkcaldy
312 Knowsley North
313 Knowsley South
314 Lancaster
315 Langbaurgh
316 Leeds Central
317 Leeds East
318 Leeds North East
319 Leeds North West
320 Leeds West
321 Leicester East
322 Leicester South
323 Leicester West
324 Leigh
325 Leominster
326 Lewes
327 Lewisham Deptford
328 Lewisham East
329 Lewisham West
330 Leyton
331 Lincoln
332 Linlithgow
333 Littleborough and Saddleworth
334 Liverpool Broadgreen
335 Liverpool Garston
336 Liverpool Mossley Hill
337 Liverpool Riverside
338 Liverpool Walton
339 Liverpool West Derby
340 Livingston
341 Llanelli
342 Loughborough
343 Ludlow
344 Luton South
345 Macclesfield
346 Maidstone
347 Makerfield
348 Manchester Blackley
349 Manchester Central
350 Manchester Gorton
351 Manchester Withington
352 Manchester Wythenshawe
353 Mansfield
354 Medway
355 Meirionnydd nant Conwy
356 Meriden
357 Merthyr Tydfil and Rhymney
358 Mid Bedfordshire
359 Mid Kent
360 Mid Norfolk
361 Mid Staffordshire
362 Mid Sussex
363 Mid Worcestershire
364 Middlesbrough
365 Midlothian
366 Milton Keynes
367 Mitcham and Morden
368 Mole Valley
369 Monklands East
370 Monklands West
371 Monmouth
372 Montgomery
373 Moray
374 Morecambe and Lunesdale
375 Morley and Leeds South
376 Motherwell North
377 Motherwell South
378 Neath
379 New Forest
380 Newark
381 Newbury
382 Newcastle-under-Lyme
383 Newcastle-upon-Tyne Central
384 Newcastle-upon-Tyne East
385 Newcastle-upon-Tyne North
386 Newham North East
387 Newham North West
388 Newham South
389 Newport East
390 Newport West
391 Normanton
392 North Bedfordshire
393 North Colchester
394 North Cornwall
395 North Devon
396 North Dorset
397 North Durham
398 North East Cambridgeshire
399 North East Derbyshire
400 North East Fife
401 North Hertfordshire
402 North Luton
403 North Norfolk
404 North Shropshire
405 North Tayside
406 North Thanet
407 North Warwickshire
408 North West Durham
409 North West Hampshire
410 North West Leicestershire
411 North West Norfolk
412 North West Surrey
413 North Wiltshire
414 Northampton North
415 Northampton South
416 Northavon
417 Norwich North
418 Norwich South
419 Norwood
420 Nottingham East
421 Nottingham North
422 Nottingham South
423 Nuneaton
424 Ogmore
425 Old Bexley and Sidcup
426 Oldham Central and Royton
427 Oldham West
428 Orkney and Shetland
429 Orpington
430 Oxford East
431 Oxford West and Abingdon
432 Paisley North
433 Paisley South
434 Peckham
435 Pembroke
436 Pendle
437 Penrith and the Borders
438 Perth and Kinross
439 Peterborough
440 Plymouth Devonport
441 Plymouth Drake
442 Plymouth Sutton
443 Pontefract and Castleford
444 Pontypridd
445 Poole
446 Portsmouth North
447 Portsmouth South
448 Preston
449 Pudsey
450 Putney
451 Ravensbourne
452 Reading East
453 Reading West
454 Redcar
455 Reigate
456 Renfrew West and Inverclyde
457 Rhondda
458 Ribble Valley
459 Richmond
460 Richmond-upon-Thames and Barnes
461 Rochdale
462 Rochford
463 Romford
464 Romsey and Waterside
465 Ross, Cromarty and Skye
466 Rossendale and Darwen
467 Rother Valley
468 Rotherham
469 Roxburgh and Berwickshire
470 Rugby and Kenilworth
471 Ruislip - Northwood
472 Rushcliffe
473 Rutland and Melton
474 Ryedale
475 Saffron Walden
476 Salford East
477 Salisbury
478 Scarborough
479 Sedgefield
480 Selby
481 Sevenoaks
482 Sheffield Attercliffe
483 Sheffield Brightside
484 Sheffield Central
485 Sheffield Hallam
486 Sheffield Heeley
487 Sheffield Hillsborough
488 Sherwood
489 Shipley
490 Shoreham
491 Shrewsbury and Atcham
492 Skipton and Ripon
493 Slough
494 Solihull
495 Somerton and Frome
496 South Colchester and Maldon
497 South Derbyshire
498 South Dorset
499 South East Cambridgeshire
500 South East Cornwall
501 South East Staffordshire
502 South Hams
503 South Norfolk
504 South Ribble
505 South Shields
506 South Staffordshire
507 South Suffolk
508 South Thanet
509 South West Bedfordshire
510 South West Cambridgeshire
511 South West Hertfordshire
512 South West Norfolk
513 South West Surrey
514 South Worcestershire
515 Southampton Itchen
516 Southampton Test
517 Southend East
518 Southend West
519 Southport
520 Southwark and Bermondsey
521 Spelthorne
522 St. Albans
523 St. Helens North
524 St. Helens South
525 St. Ives
526 Stafford
527 Staffordshire Moorlands
528 Stalybridge and Hyde
529 Stamford and Spalding
530 Stevenage
531 Stirling
532 Stockport
533 Stockton North
534 Stockton South
535 Stoke-on-Trent Central
536 Stoke-on-Trent North
537 Stoke-on-Trent South
538 Stratford-on-Avon
539 Strathkelvin and Bearsden
540 Streatham
541 Stretford
542 Stroud
543 Suffolk Coastal
544 Sunderland North
545 Sunderland South
546 Surbiton
547 Sutton and Cheam
548 Sutton Coldfield
549 Swansea East
550 Swansea West
551 Swindon
552 Tatton
553 Taunton
554 Teignbridge
555 The City of London and Westminster South
556 The Wrekin
557 Thurrock
558 Tiverton
559 Tonbridge and Malling
560 Tooting
561 Torbay
562 Torfaen
563 Torridge and West Devon
564 Tottenham
565 Truro
566 Tunbridge Wells
567 Tweeddale, Ettrick and Lauderdale
568 Twickenham
569 Tyne Bridge
570 Tynemouth
571 Upminster
572 Uxbridge
573 Vale of Glamorgan
574 Vauxhall
575 Wakefield
576 Wallasey
577 Wallsend
578 Walsall North
579 Walsall South
580 Walthamstow
581 Wansbeck
582 Wansdyke
583 Wanstead and Woodford
584 Wantage
585 Warley East
586 Warley West
587 Warrington North
588 Warrington South
589 Warwick and Leamington
590 Watford
591 Waveney
592 Wealden
593 Wellingborough
594 Wells
595 Welwyn Hatfield
596 Wentworth
597 West Bromwich East
598 West Bromwich West
599 West Derbyshire
600 West Dorset
601 West Gloucestershire
602 West Hertfordshire
603 West Lancashire
604 Westbury
605 Western Isles
606 Westminster North
607 Westmorland and Lonsdale
608 Weston-Super-Mare
609 Wigan
610 Wimbledon
611 Winchester
612 Windsor and Maidenhead
613 Wirral South
614 Wirral West
615 Witney
616 Woking
617 Wokingham
618 Wolverhampton North East
619 Wolverhampton South East
620 Wolverhampton South West
621 Woodspring
622 Woolwich
623 Worcester
624 Workington
625 Worsley
626 Worthing
627 Wrexham
628 Wycombe
629 Wyre
630 Wyre Forest
631 Yeovil
632 Ynys Mon
633 York

Figure 3.9 The official names of all the 633 constituencies that existed in 1981 are shown, sorted and indexed alphabetically. Places can be identified and individual constituencies found on the previous map and in the next cartogram using this list. The print is very small as there are so many areas with such long names. A magnifying glass can help in reading them.

Parliamentary Constituencies

(Equal Area Cartogram)

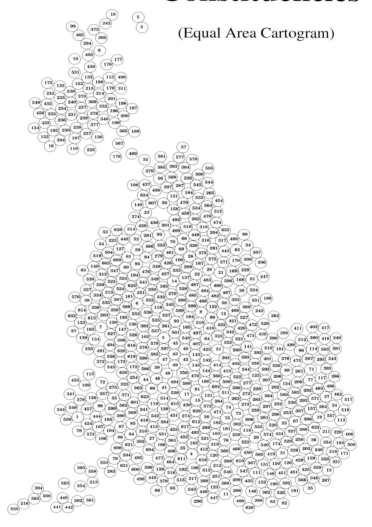

Figure 3.10   The 633 constituencies are labelled for identification against an alphabetical list. Their size is based, in this case, on the total 1981 census population. Although some places do differ significantly in population size, North and South Aberdeen, for instance (numbers 2 and 3), there is little pattern to this size variation.

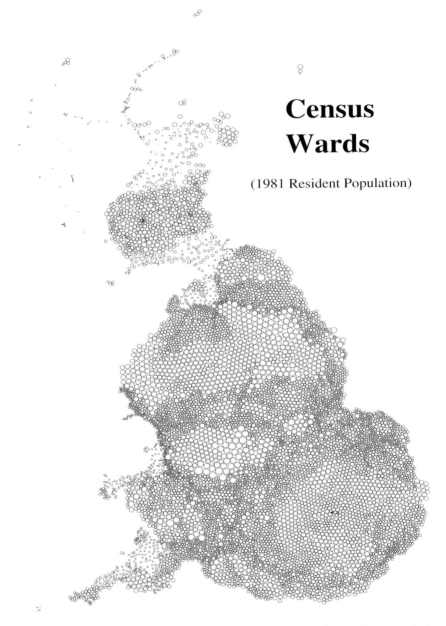

Census Wards

(1981 Resident Population)

Figure 3.11   The area of the hexagons is in proportion to the resident population of 10 444 1981 census wards. Birmingham has the largest wards and the smallest are within the City of London, creating patterns. In Scotland the hexagons are part-post-code sectors. Some contain few people, particularly in central Glasgow and Edinburgh where there were many businesses, but few people slept.

created, appear continuous again – building Thiessen polygons around the circle centres is simplest. Here every straight edge side of a polygon is equidistant from the two nearest circle centres and hence every vertex is equidistant from the three nearest circle centres.

With a circular cartogram we can now create a space of places and a space that maintains, as far as is possible with circles, the original topology.[14] Such a cartogram can be particularly useful for certain types of visualization, those involving thousands of areas, as it presents a much clearer image than one that would have to twist and wind to satisfy strictly all the topology conditions, all of the time.

3.5   Producing illusions

> *So gravity can be explained by assuming that matter curves space. But why should matter do this? Why should matter curve space? One explanation is that space curvature is what matter is. William K. Clifford first proposed this theory in an 1870 paper called 'On the Space Theory of Matter'.*
>
> *(Rucker, 1984, p. 82)*

In this book the density of people is used to curve the space they live in. What places should now be chosen out of which to build these abstract spaces? How will the choice of which hierarchy and division of areas to use alter the image (Figures 3.12 and 3.13)? Many different cartograms, each based upon the same population head count but drawn separately from all the major administrative divisions of Britain, have been constructed to show how much the division of space matters, and many are shown in these pages.

The answer to the question of robustness is that the choice of an areal unit does not substantially alter the final shape of these images – this is a reassuring outcome. In fact all thoughtfully constructed cartograms of Britain (or any other territory) tend towards the same rough structure, which loosely implies that an ideal solution exists. There is also a sense of aesthetic acceptance to be realised.[15]

---

[14] The cartograms used in this work are pseudo-continuous: 'There are two types of contiguous cartograms. The first is the pseudo-continuous cartogram. ... Pseudo-continuous cartograms depict regions like a continuous map, but are endowed with the "pseudo" label (after Muehrke, 1978) due to the generalization of the polygon's topological structure. This can be contrasted with the contiguous cartogram, where the topology has been retained. ... ' (Torguson, 1990, p. 17).

[15] The following papers document various attempts to control or automate the process: Hunter and Young (1968), Hunter and Meade (1971), Skoda and Robertson (1972), Tobler (1959, 1973a, 1986), Olson (1976), Kadmon and Shlomi (1978), Eastman, Nelson and Shields (1981), Dougenik, Niemeyer and Chrisman (1983), Nelson and McGregor (1983), Cuff, Pawling and Blair (1984), Selvin *et al*. (1984), Dougenik, Chrisman and Niemeyer (1985), Kelly (1987), Cauvin, Schneider and Cherrier (1989), Torguson (1990) and Dorling, Barford and Newman (2006).

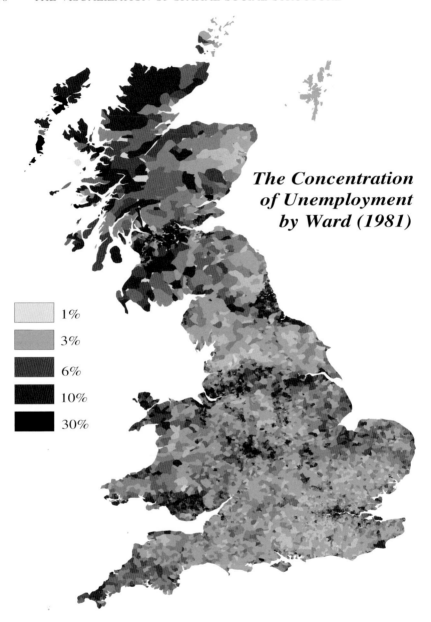

Figure 3.12   The 10 444 wards showing unemployment rates from the 1981 census on an equal land area map using a continuous grey scale. This unsuitable projection dramatically overemphasises the high rates of unemployment in the underpopulated North West of Scotland, where the fate of less than a dozen people could darken a large expanse of land. The inner city problems can just be made out.

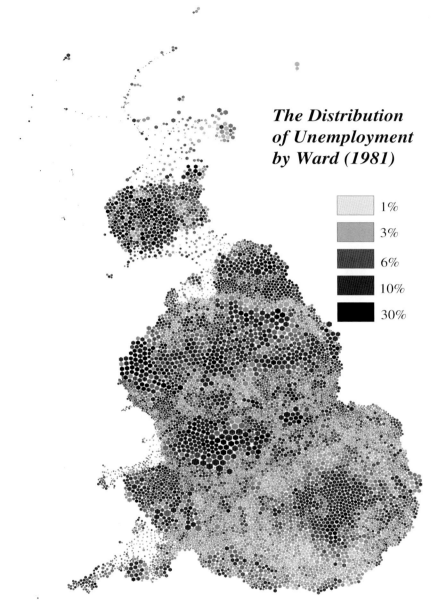

**The Distribution
of Unemployment
by Ward (1981)**

1%
3%
6%
10%
30%

Figure 3.13   Important areas with high rates that can be identified including the Welsh Valleys, West Midlands, Liverpool, South Yorkshire, the North East conurbations and the Glasgow area. The small cluster of darkly coloured wards making up Corby can be seen just to the right of centre – now to scale. Areas with very low rates of unemployment form a distinctive pattern.

The degree of autonomy that the final shape of the cartogram has, from the influence of the areal division that was used to create it, was only achieved by choosing a careful definition and measure of *contiguity*. In creating the cartograms shown here two places were said to be *contiguous* if they shared a common border or were linked by a major tunnel, road or rail bridge. The measure of contiguity was not absolute, but estimated as the proportion of the perimeter of an area made up by the border in question or the length of an estuary coastline, which a bridge, for instance, rendered traversable.

The algorithm for creating the cartograms began with each place at its original (Euclidean) location, represented as a circle whose area was in proportion to its population. Next, overlapping circles repelled each other while circles were attracted to their neighbours in relation to the strength of their contiguity measure. Places that bordered the sea expressed a degree of inertia because part of their perimeter, being coastline, did not make up a common border, and this helped to maintain prominent peninsulas. Thus, although the exact shape of the coastline was sacrificed, many of its key locational features were retained.

The sustained combination of all these forces in parallel (Box 3.3) created the new pictures of Britain used throughout this book.[16] An algorithm was written where the solution evolved towards the desired goal, releasing and tightening

Box 3.3   Deriving a constant

The following set of equations show how the damping factor K of 0.25 was derived for the cartogram algorithm, where x is position, v is velocity and A^n is used as an *ansatz*:

$$v_{n+1} = K(v_n - x_n)$$

$$x_{n+1} = x_n + v_{n+1}$$

$$x_{n+1} = x_n + K((x_n - x_{n-1}) - x_n)$$

$$x_{n+1} - x_n + K x_{n-1} = 0$$

$$A^{n+2} - A^{n+1} + K A^n = 0$$

by factors:

$$\text{either} \qquad A^n = 0$$

$$\text{or} \quad A^2 - A + K = 0$$

To avoid oscillation the solutions to A must be real; therefore the discriminant must be positive: $1 - 4K \geq 0$. Thus the largest nonoscillating K is 0.25.

---

[16] These were also used in 'A New Social Atlas of Britain', published in 1995 and made available as an open access 131 Mb pdf file on the web: http://www.dannydorling.org/?page_id = 81.

constraints to allow the conditions to be attained and to ensure that the final pictures looked acceptable.[17]

3.6   Population space

It is not miles or kilometers across the surface of the earth in which these geo-political factors are arranged, but rather an interesting – but complicated – series of topological relationships.

(McCleary, 1988, p. 148)

The very shape and layout of the cartogram is of interest even before we begin to use it to depict other information. The population cartogram tells us a lot about the human geography of places – how they are related to each other in a new and intriguingly unfamiliar way (Figures 3.14 and 3.15).

The population of Britain is more drastically dominated by London than even most human geographers would imagine. Greater London itself contains over an eighth of the population. If Greater London is combined with those areas under London's immediate influence in population space, we can count nearly half the people of the island.

The areas of influence of the other great cities are clearly shown, as is the way they compare and combine, are divided and divide space up among themselves (Figure 3.16). The separation of Wales into North and South, and Scotland from England, highlights divisions that are well known, but missing from conventional depictions.[18]

To make the reading of the cartogram simpler and to learn more about population space, we can transform the major networks of infrastructure, which service the population and along which they move, to lie upon the transformed space. The layout and purpose of the mainline railway network (Figure 3.17) is clear on the cartogram (Figure 3.18). It provides a series of arteries attempting to reach all areas equitably, in accordance to their populations.

The road network of Britain is much more complex, and only the motorways and designated main routes are shown in the illustrations drawn

---

[17] The algorithm was published in Dorling (1996), a copy of which can be found here on the web: http://qmrg.org.uk/files/2008/11/59-area-cartograms.pdf. Also see the website: http://www.ncgia.ucsb.edu/projects/Cartogram_Central/ for more cartogram ephemera.

[18] The use of cartograms makes some very simple mapping possible: 'Absolute numbers should never be mapped ... when using a standard set of boundaries. To do so is grossly misleading since large areas will automatically tend to be black. Two solutions exist:

– to standardize the data, most commonly by converting the variable to a percentage or other ratio form;

– to transform the map base, such that the basic areas are enlarged or reduced in size so as to represent the total numbers of people therein, then to map, say, absolute numbers of retired people on this new base map. ...' (Rhind, 1983, p. 187).

Figure 3.14   The 64 counties are shaded so that no areas with the same colour are contiguous. This map can be used as a key to the next enumeration district cartogram shown in this book, where the very small areas are each coloured the same shade as their respective counties. In Scotland these areas were called 'regions' in the 1980s, the old counties having become defunct.

Enumeration District Cartogram

(Counties and Inner London
Shown by Grey-Scale)

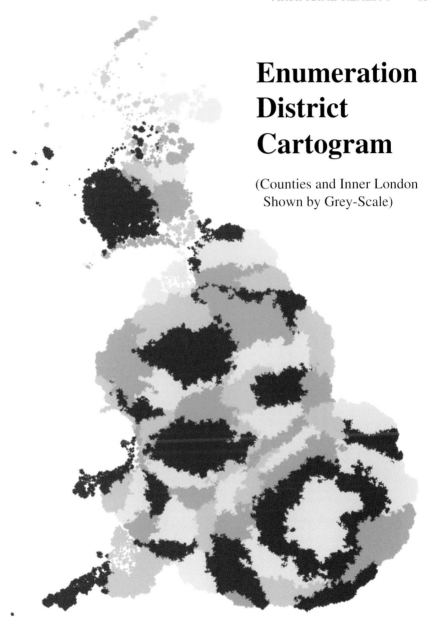

*Figure 3.15   The cartogram is based on the 1981 census resident populations of
129 211 enumeration districts. Each district is coloured according to the county
to which it belongs with the same shade as on the previous map. Inner London
has been added as a light shade of grey. Note that the county borders take on a
fractal pattern. Lines that are simple in one metric can be complex in others.*

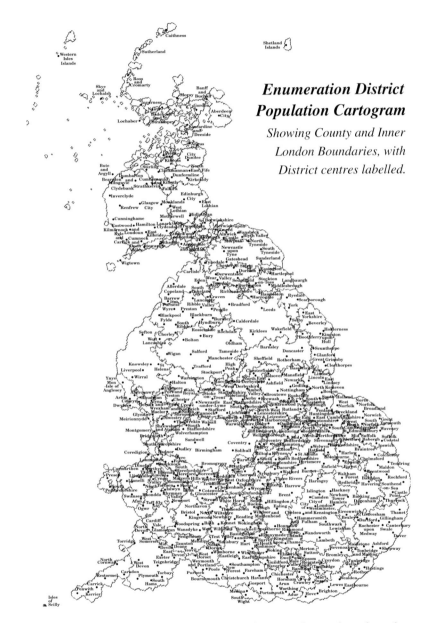

Enumeration District
Population Cartogram

Showing County and Inner
London Boundaries, with
District centres labelled.

Figure 3.16   *This cartogram is the same as that just shown, based on the same population of the 1981 census enumeration districts. County and Inner London boundaries are shown and district centres labelled. The district names have been rearranged by hand after application of a boundary edge-detection algorithm. This diagram can be used as a key to the enumeration district cartograms shown later.*

here (Figure 3.19). Again the even spread across the country can be noted (Figure 3.20). Intriguingly, though, the network is most sparse in population space where it is most concentrated on the ordinary map – in London. It is no wonder that congestion is greatest where there are least roads per head of population.

3.7   Stretching spacetime

A linear cartogram operates like an area cartogram but instead of varying areas with values it is the map distances which vary with values. Its construction is analogous to the azimuthal equidistant projection, but rather than physical distance varying with map distance – cost of effort of travel are used.

(Lai, 1983, p. 33)

The difficulty of constructing area cartograms is due to the fact that they are two-dimensional entities (Figures 3.21 and 3.22). One-dimensional cartograms are simplicity itself to produce.

Imagine a one-dimensional, temporal cartogram of the human population of the world ever born, from when the species began until the present day. Such a cartogram would consist of a single line, with dates marked along its length (Box 3.4). The distance between any two dates would be in proportion to the number of people born between those times. Thus, the time line would be very compact at the beginning, having its years widely spread towards the end. More importantly, it is unambiguously the only solution to the problem of including all people who have ever lived.

The number of dimensions of a cartogram can vary, limited in type only by the imagination. A halfway house can be envisaged of a one-and-a-half-dimensional cartogram, where some information independent of time is depicted vertically up from the time-line cartogram, for example the proportion of the population living in the various continents. Such a cartogram would be just as simple to construct and, while appearing two-dimensional, the information is of one dimension (place) within another (time). The second graph in Box 3.4 shows this.

The term linear has already been reserved in the literature on cartograms to mean something other than one-dimensional. The most well-known reason to draw these cartograms has been to fit place names in and so simplify the mapping of a city's underground train system. Another well-known option is to make the distances between places proportional to the time or cost required to travel between them.[19] This can only be achieved for the, say, shortest travel

---

[19] 'The plotting of places in terms of accessibility metrics like time and cost distances is particularly valuable when communicable diseases are being studied and may frequently provide a fresh perspective on the disease patterns occurring' (Cliff and Haggett, 1988, p. 267).

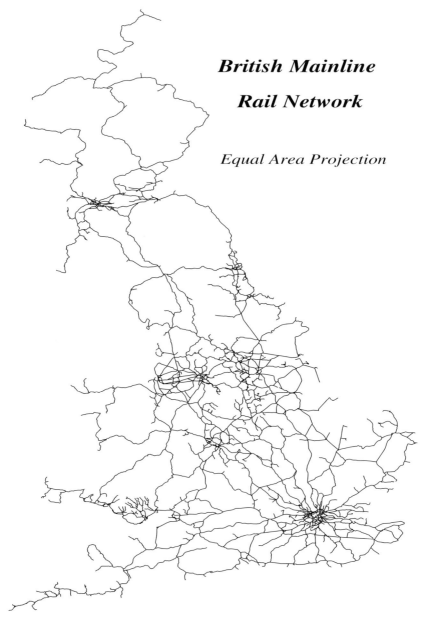

British Mainline Rail Network

Equal Area Projection

Figure 3.17   Railway tracks in 1980 digitised from the 1:625 000 ordinance survey maps. The rail network appears, at fist glance, to be very concentrated. It certainly avoids high land – but then, so do people. This is the network that was left after the cuts to services, cuts that took place mainly in the 1960s, but also before and after then. The rail network was most extensive in 1913.

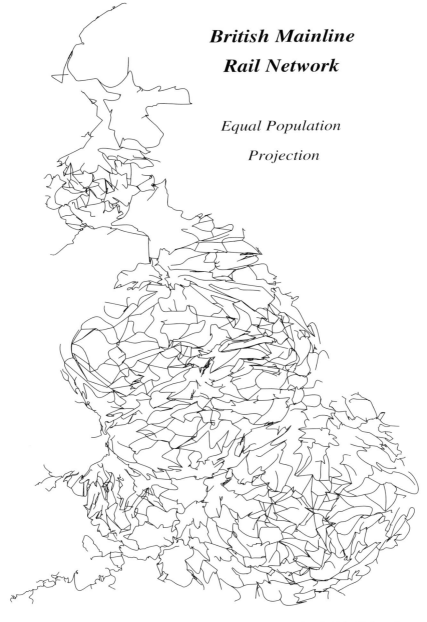

**British Mainline
Rail Network**

Equal Population

Projection

Figure 3.18   The 1980 railway tracks digitised from the 1:625 000 ordinance survey maps and shown on an equal population projection. The rail network can be seen to cover the population very evenly, despite the fact that most of it was laid down over one hundred years earlier. Probably the cuts to services made the distribution more even as it was mainly remoter rural services that were cut.

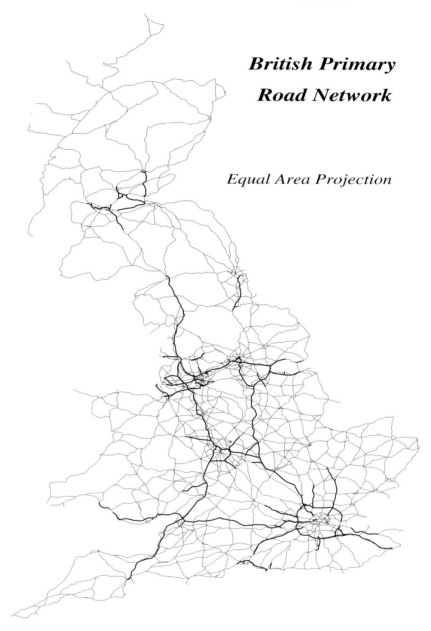

**British Primary
Road Network**

Equal Area Projection

Figure 3.19   The 1980 motorway and trunk routes taken from the 1:625 000 ordinance survey digital records. The motorway network (in thicker lines) can be seen to link the major conurbations. The trunk road network covers the space of physical geography fairly evenly. The two sets of roads look a little like major and minor arteries serving particular urban areas most densely.

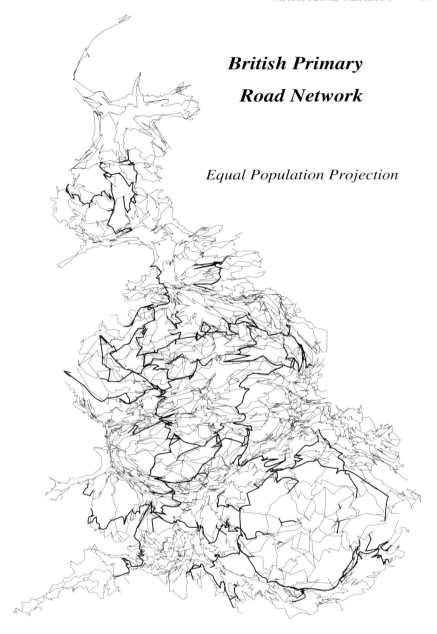

**British Primary
Road Network**

Equal Population Projection

Figure 3.20   The motorways are the thicker lines drawn on this equal population cartogram. Roads are most concentrated, per capita, in more rural areas. The cartogram is based on night-time population, so if it were redrawn using daytime figures all the cities would be very much larger. London, in particular, would appear to be very much more poorly served with roads. It has the least space.

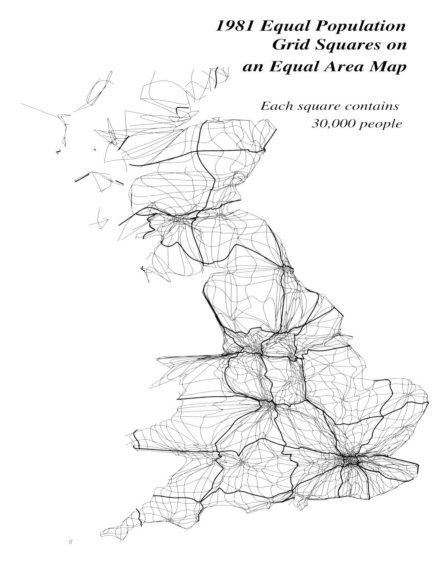

1981 Equal Population Grid Squares on an Equal Area Map

Each square contains 30,000 people

Figure 3.21 This map was created using enumeration district populations. The darker lines encompass areas of three million people and 100 smaller 'squares'. If you did want to map population grid squares on a normal map, perhaps this lattice would be most appropriate. London, as expected, contains the mostly densely packed single large population square. That square is home to three million people.

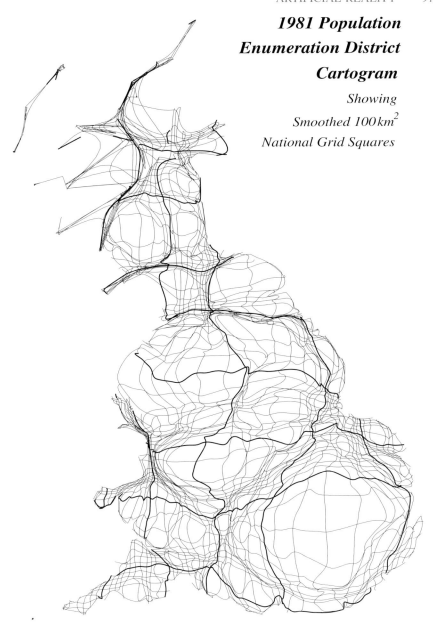

**1981 Population
Enumeration District
Cartogram**
Showing
Smoothed $100\,km^2$
National Grid Squares

*Figure 3.22   National grid lines are drawn on this population cartogram at
10 kilometre intervals, thicker every 100 kilometres. The lines were statistically
smoothed. The distortion of the London area is dramatically depicted. It is almost
as if the cities are rising out of the paper. Note that all lines should meet at right
angles - approximating what is known as conformality.*

Box 3.4   Many-dimensional cartograms

Many different types of area cartogram can be imagined. Here are some world scale examples:

One-dimensional: World Population over time

One-and-a-half-dimensional: World
   Population over time by continent

Two-dimensional: World Population over space
   by continent

Two-and-a-half-dimensional: World
   Population with income as
   perceived height

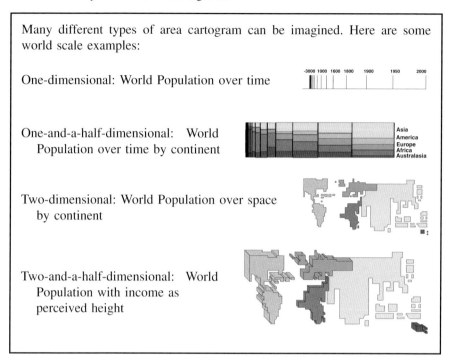

time distances between all places, when the two-dimensional space in which the linear cartogram resides is itself warped in the third dimension.

What happens when we go beyond the two spatial dimensions and also attempt to incorporate time? At one level an analogue to the one-and-a-half-dimensional cartogram can be made. The linear cartogram where distance is made proportional to travel time can be projected as a surface above an area cartogram. Thus a two-and-a-half-dimensional linear area cartogram is created as a surface of travel time above population space.[20]

Even simple two-dimensional population space changes in time, dramatically so over long periods. A series of area cartograms has been constructed for this book of the British electorate by parliamentary constituency from 1955 to 1987. The ten images show the gradual deformation of the space as the electorate grows nationally, the South East swelling in particular while the inner cities shrink. The fact that the definition and number of places changed also over this period was easily incorporated in the graphics.

---

[20] Angel and Hyman (1972, 1976). Note that Section 7.6 in this book gives more details.

A true three-dimensional volume cartogram of population spacetime is difficult to imagine.[21] Such an image would have to be based upon the axiom of giving each human life equal representation rather than each area. As lives have temporal extent they would have to be drawn as life lines. It is hard to imagine what further constraints would be employed in constructing such spaces. Obviously volume should be in proportion to individual lives and contiguous places in space should touch each other, as should places connected with themselves, both forwards and backwards in time.

If we then choose to minimise the area of internal boundaries, which in a three-dimensional spacetime cartogram are planes rather than lines, we will warp time into space and vice versa.[22] A place that many people left will slip back in time, while a place growing in population pushes forward. What are we creating and how can we understand it – let alone view it?

The computer algorithm employed here could be adapted to create all the variants mentioned above. The problem is not creating them; it is understanding why it would be useful to do so and how to use the transformed manifolds made. The nature, creation and use of spaces above two dimensions form the subject of the last part of this book. Here, next, concentration is applied to how the unusual, but understandable, two-dimensional population spaces can be gainfully employed in the visualization of spatial social structure.

---

[21] Two-dimensional cartograms of death are hard enough to envisage, let alone three-dimensional ones of birth, life and death: 'Turning to the map of the world, how far from London can you extend your military power before you will lose one hundred thousand men? This "circle" obviously moves much farther over sea than into Europe where military resistance, say from the French, would make such a move very expensive per geographic mile. A map of the earth with hundred-thousand-man-lost circles centred on England in 1850 establishes the most distant place, not as New Zealand, but somewhere around Moscow. The paths of least deaths at right angles to the circles draw another set of real longitudes and latitudes. Moscow is antipode, the opposite side of the earth, the "down under" from London. To conquer the world is to conquer Moscow, not Auckland, and this is why the British kept moving toward Moscow from the Crimea, the walls of Peking, the Khyber Pass, from Vladivostok. All "paths" from London lead to Moscow, and thus the mysticism of Mackinder's geopolitic is explained' (Bunge, 1973, p. 286).

[22] 'In the opinion of this author, the value of such maps can be great for geographers and other behavioural scientists – a value which seems limited only by the imagination of the scholars whose tools they should be' (Lewis, 1969, p. 406). Section 9.3 of this book explores these ideas further.

4

Honeycomb structure

Detail cumulates into larger coherent structures; those thousands of tiny windows, when seen at a distance, gray into surfaces to form a whole building. Simplicity of reading derives from the context of detailed and complex information, properly arranged. A most unconventional design strategy is revealed: to clarify, add detail.

(Tufte, 1990, p. 37)

4.1 Viewing society

Britain is a small piece of land upon which, by the early years of the twenty-first century, not much more than sixty million people lived. It was probably about five million fewer when the first draft of this text was written. The five million is a net change. Far more were born, far more died, far more came and far more went from Britain's shores than five million. Five million is just the difference of the difference between these numbers. It is births less deaths, plus ins less outs (Figure 4.1).

Sixty million is a high number compared with most other European countries. It is a low number compared to two other large countries in which many people speak English: India and the USA. Why study this net number, less than a single percentage of the world's population? Why study what it looked like back when it ranged between 55 and 56 million, between the spring 1971 census and autumn 1989 housing crash?

The British state is a convenient unit of analysis; consistent national statistics are collected on many subjects at regular intervals about all its constituent

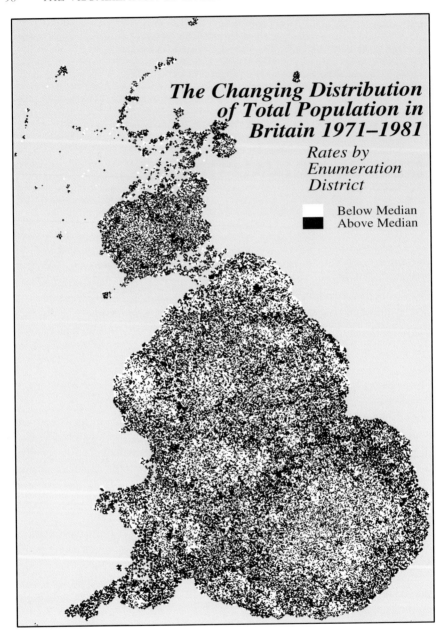

Figure 4.1   Black or white shading of the 129 211 1981 census enumeration districts using an equal population cartogram. Black areas are growing relative to average change and are all growing in terms of absolute numbers. White areas are of relative population decline; in most cases this is an absolute population decline. The drift from the conurbations to country can be seen if you squint.

elements – often to be mapped.[1] If Northern Ireland is added, a single political body is created from areas covering all its territory, and its territory is divided into areas that are themselves subdivided into units of local representation for political election. Apart from the border with Ireland, in the 1980s its boundary was well defined, and encompassed almost all the movements of people who live within the state borders. Today the border with Scotland is becoming disputed as Scottish independence is debated more widely.

The British state is an understandable social entity. The definitions of people, work, places, political parties and groups are simple. Its history and geography are well documented and easily accessible. It covers too little land area and too many people to divide easily into smaller wholes that make sense to study alone, even if they are separate countries. Its main land mass can be travelled from one end to the other, by land, in a day; but only a few of its people can be met in a lifetime.

4.2   Who the people are

> ... *superimpose all the characteristics on the same map.* ... *But then the question: where is a given characteristic? no longer has a visual answer. Should this question indeed have an answer? This is the basic problem in cartography with* n *characteristics, that is, 'thematic' or more precisely, 'polythematic' cartography.*
>
> *(Bertin, 1981, p. 140)*

The pictures drawn to accompany this chapter are mainly based on population cartograms of one hundred and thirty thousand enumeration districts[2] (Box 4.1). This resolution was chosen as the finest that is possible – closest to the local scale and individual realities of life. Great regional patterns can still be seen in the images, but only where they really exist, not as fabrications of the boundaries chosen. We have collected a greater quantity of routine information on people in

---

[1] 'The graphic portrayal of census data has always been a decentralised and in many respects an *ad hoc* affair. After the superb maps produced by Petermann (partly for the government) after the 1841 and 1851 censuses, little "official" mapping was done until that carried out after the 1961 Census, by what is now the Department of the Environment (DOE). A tradition grew up that individual geographers mapped those elements of the census in which they were interested and in 1968 one of the Transactions of the Institute of British Geographers consisted of a set of twelve maps of variables from the 1961 data. This 7-year delay in map availability was very similar to that after the 1841 census' (Rhind, 1975b, p. 9).

[2] Box 4.1 explains how the 1971 Census was compressed to two files of 29.9 Mb and 11.3 Mb in size to allow the data to be accessed quickly. The entire 1981 Census was similarly compressed to a few files that now appear relatively small in size. Compressed like this, all the British censuses ever digitised could now be fitted on a cheap pen-sized 32 Gb data stick, but we no longer need to be so frugal with file space. Later in the chapter Box 14 explains how data for the 125 thousand enumeration districts of 1971 were linked to the 130 thousand of 1981 so that both censuses could be mapped using the latter areas.

Box 4.1   Storing the census

The small area statistics of the 1971 whole population census consisted of 480 counts for each of 125 476 enumeration districts, some sixty million numbers. Originally each count was stored in an eight character wide slot and the file was half a gigabyte in size, far too large to be easily stored and repeatedly accessed. A simple form of run length encoding was customised to compress the file and still allow the records of individual enumeration districts to be read instantly. The counts for each cell were stored sequentially as either a run of zeros, half-bytes (for 0 to 15), bytes (for 0 to 255) or half-words (for 0 to 65 535). The sophistication of the algorithm was in deciding when it was profitable to drop down an order of magnitude in the form of storage used and when it was not. This was achieved by looking through the list both forwards and backwards. The following simplified heuristic was employed:

define: yesterday, today and tomorrow as the magnitude of the previous, present and future cell to be encoded. Then, if the opportunity to lower the magnitude of storage arises (today<yesterday) continue at the present order while tomorrow ≥yesterday.

With a few other caveats, this rule compresses the file to just 5 % of its former volume: under 30 megabytes. The more sparse first section, the 10 % population census file containing 368 cells by 125 462 enumeration districts (14 missing) is compressed to a file of just 11 megabytes in size. These figures are better than those achieved by the standard Lempel-Ziv compress algorithm, but, more importantly, the file could be read and decoded faster than any other configuration (including the original flat form) given disk speed restrictions in 1990.

the last twenty years than over the previous twenty thousand. Is it not surprising that radically new techniques are required to view the social landscape?[3] Conventional choropleth maps at the level of ten thousand wards are occasionally included here to show how they contrast with the message of the cartograms.

Gender is the least ambiguous attribute we give people. One of the very first cartograms made for this book (Figure 4.2) of enumeration districts was a picture where each street block is coloured either black, for over-average proportions of females, or white, for under-average. The picture not only showed the random variation in this statistic, indicated by the speckled nature of the image, but also

---

[3] 'Quite simply there is far too much information to allow policy-makers, planners, geographers, politicians, schoolchildren, and others interested in census data for a particular area to be able to identify easily patterns of characteristics or features of interest from SAS [Small Area Statistics] data without processing and condensing it in some way' (Openshaw, 1983, p. 243).

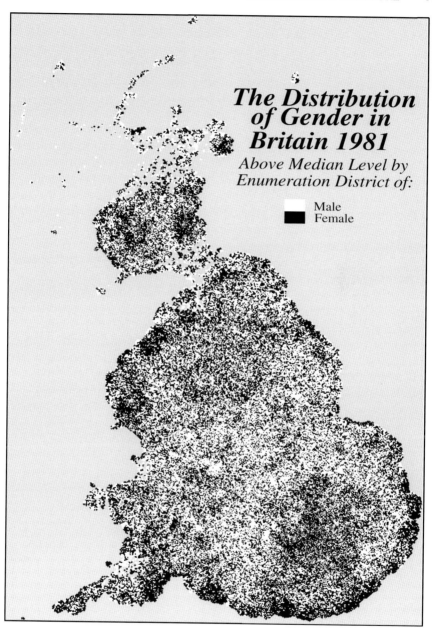

The Distribution of Gender in Britain 1981

Above Median Level by Enumeration District of:

Male
Female

Figure 4.2   Areas are shaded black where above median proportions of women and girls lived. In all cases this means majority female areas. The coasts, much of central London and some other (especially Scottish) cities are seen to be home to more women than men. The white areas are where there are relatively more men. The Home Counties ring of areas with more men is distinctive.

suggests simple patterns of slight over- and under-representation. Contrast the speckled image of male and female distribution with the much more structured patterns by age (Figure 4.3). There are more women in the middle of cities and along the South coast (Figures 4.4, 4.5 and 4.6). A similar two colour technique is used later to show the distribution of Irish born, where the patterns are more clear (Figure 4.7).

To see distinctly the geographical distribution of the proportions being mapped, say the proportion of all people in an area who are women, a relative scale of measurement is adopted – above or below the median level of females rather than above or below fifty per cent. Thus half the area of Britain, on the population cartogram, is shaded black, the other white. In this example the picture would hardly alter if an absolute scale were used. However, the image for the Irish born would look very different if only areas where a majority of people were born in Ireland were coloured black, or even some much lower proportion. Absolute scales would require different legends for all pictures. The images would also vary greatly in their levels of saturation, merely because of the use of arbitrary, not comparable, cut-offs.

Here, levels above and below the median for Britain or, in some cases, groups bounded by quartile levels, are used to shade the images. This is done to treat all issues and hence variables most simply, consistently and comparatively. The division into two levels, quartiles and beyond, can be extended until continuous shading is achieved. This has not been used here at enumeration district level, as it is difficult to shade, or see, such small areas continuously. More complex colouring schemes are developed later, but, as the images stand, continuous impressions are gained through the dithered patterns created by so many tiny discrete shades.[4]

The main influence upon the patterns shown by the distribution of gender is age. Women tend to live longer, so where the population is generally older it is likely to contain more women. The distribution of the elderly, in every neighbourhood in Britain, can also be shown, in the same way as the distribution of women was depicted. The two maps could be compared, but shortly we will see how both variables can be shown on a single map, with yet a third variable, the distribution of the young, introduced through the use of colour mixing.

A traditional means of showing information about the contrasting demography of locations is to draw numerous population pyramids, but these fail to convey the distribution of age and sex structure across more than a few large geographical areas. Pyramids are, in addition, not easily visually comparable. However, some profitable use is made of symbols similar to the pyramid shown later in Chapter 8.

---

[4] The use of many small units can be repeatedly justified from the errors that manifest when they are not employed: 'An odd consequence of the redrawing of county boundaries in 1974 was that Lancashire became an area of concentration of the elderly' (Warnes and Law, 1984, p. 40).

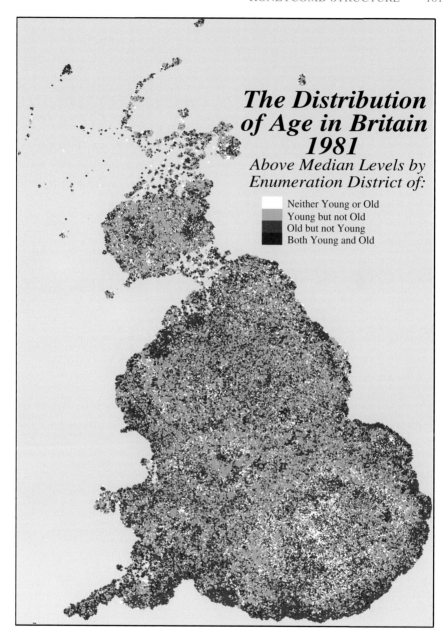

The Distribution of Age in Britain 1981
Above Median Levels by Enumeration District of:

- Neither Young or Old
- Young but not Old
- Old but not Young
- Both Young and Old

Figure 4.3   Enumeration districts are coloured white if there were unusually high numbers of working age people below retirement age in them and also few young or old folk. There were a few such areas in 1981 and were mainly found in London. There were also a few areas with both more old and more young people. Younger families lived away from the coasts and from some parts of London.

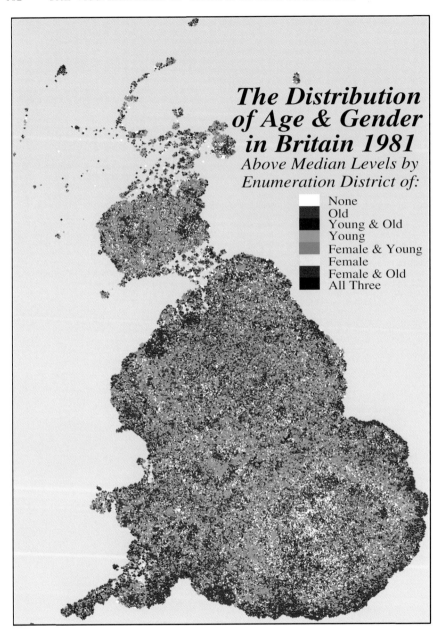

Figure 4.4   This picture is dominated by reds or cyan despite eight categories being possible. Enumeration districts tended to either have a more young, male, population or were more elderly with more females. This is only partly due to women living longer. Other colours are extremely rare. The yellows in central London are where there are more women, but not more children or more elderly.

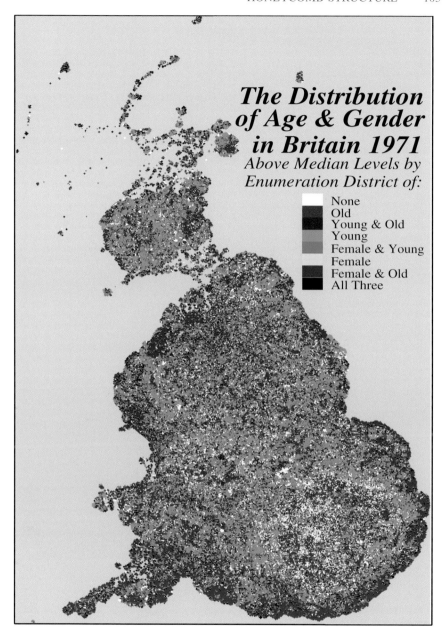

*The Distribution
of Age & Gender
in Britain 1971*
*Above Median Levels by
Enumeration District of:*

None
Old
Young & Old
Young
Female & Young
Female
Female & Old
All Three

Figure 4.5   The same three binary variables are shown here as were shown in the previous figure, but ten years earlier. There were, perhaps, a few more clusters of black areas containing more women, the young and the old, which means fewer adults aged 16 to 64 years and fewer males. Had some men left these areas, but returned by ten years later? Glasgow was also a little younger in 1971 than in 1981.

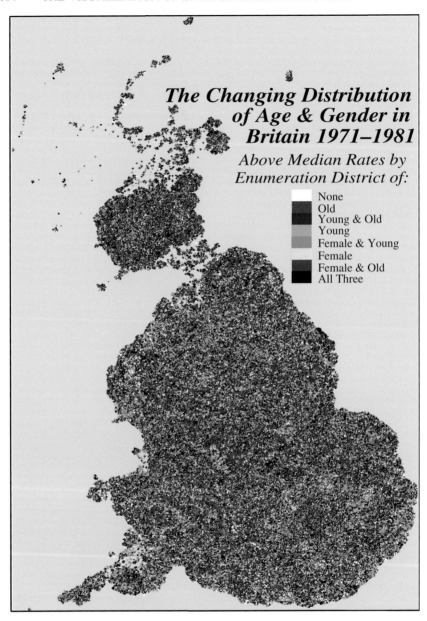

Figure 4.6   It is easier to see demographic change if that change is shown on a single image rather than two images, to be compared. Here, suburban London saw a slight influx of young women as did much of the south coast (partly care workers and the elderly people many of them cared for). Squint and you can also see a tendency for older people and men to be moving away from city centres.

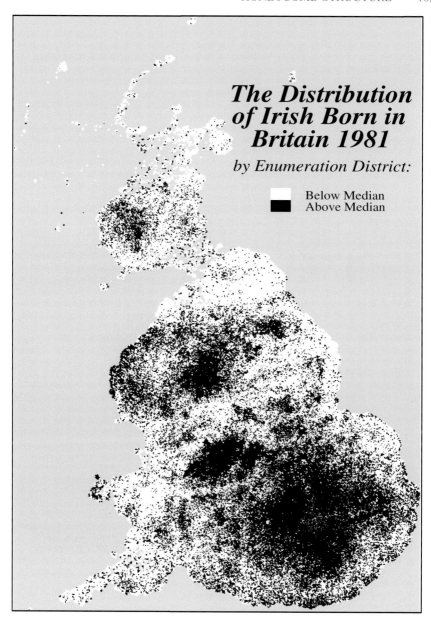

The Distribution of Irish Born in Britain 1981

by Enumeration District:

Below Median
Above Median

Figure 4.7   Black and white shading of all 129 211 census enumeration districts on an equal population cartogram. The geographical concentration of those born in Ireland is very clear: London (but not along the river!), Birmingham and Manchester dominate, followed by Leeds, Liverpool and Glasgow. By using the median level, exactly half the tiny enumeration districts are shown black.

4.3   Disparate origins

Only four pieces of information were collected about each person in the 1841 Census. That one of these was birthplace is indicative of how essential this item was and still is.

(Craig, 1987, p. 33)

Where the people who make up the social landscape of Britain came from in the past is, perhaps, as well known as it will ever be. Many people know little of their actual origins or even things as simple as the fact that most ancient Britons at one time spoke a language similar to Welsh. Researchers have sometimes been biased towards seeing some immigrant groups as more immigrant than others[5] (Figure 4.8).

Visualizing migration streams is one of the major themes in this book. Static pictures of migration are best provided by looking at the distribution of people across the country born in a particular place. Shading every street in Britain by the proportion of its population whose birthplace was in Ireland (North and South) shows the scattering of people who, in the course of their lifetime, flowed from that one island to live in this. We see immediately how strongly the Irish immigrants are concentrated in particular localities (Figure 4.9).

Migration is about mixing. The picture fails if it does not convey the colourful mixtures of people that result from their movement. Colour, resulting from the mixing of light, gives the clearest images of the kaleidoscope of people's differing origins. Not more than three primary colours can be used. That is the most that human eyes can distinguish.

Each country naturally contains mostly people born within its boundaries. Taking colours from national flowers: England is red, Scotland blue and Wales yellow. Mixing takes place between them: thus the Scottish border is purple, the Welsh border (and Cornwall) orange. There are disproportionate numbers of Welsh and Scottish people in London, where there is a dearth of English born, colouring the Capital green (a mixture of over-average amounts of blue and yellow).

Closer inspection shows just how intricate the pattern of mixing is. The white areas on the picture of the distribution of British-born are made up of streets where there are shortfalls of all the indigenous nationalities. Here overseas-born immigrant populations are most densely settled in the social landscape.[6]

---

[5] 'The emphasis of immigrant community research by British geographers and other social scientists during the past twenty years has been overwhelmingly on the Afro-Caribbean and Asian groups at the expense of those of longer standing and greater numbers, but perhaps of less visibility' (King and Shuttleworth, 1989, p. 64).

[6] It must be remembered that we are mapping place of birth, not colour of skin: 'For example, of the 322 670 persons born in India living in Britain in 1971, between one-fifth and one-third (66 139–104 362) may have been Whites born in India...' (Peach, 1982, p. 24).

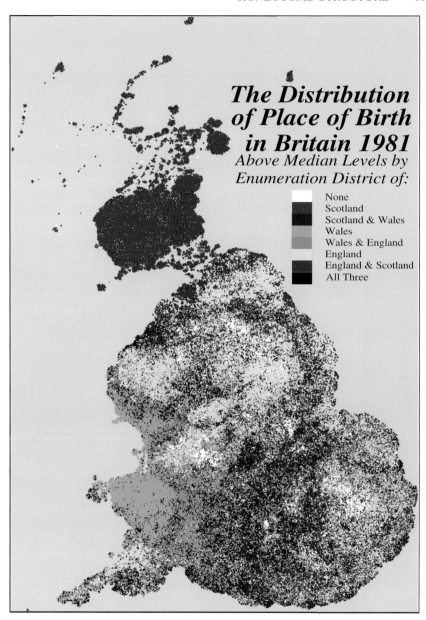

**The Distribution
of Place of Birth
in Britain 1981**
*Above Median Levels by
Enumeration District of:*

None
Scotland
Scotland & Wales
Wales
Wales & England
England
England & Scotland
All Three

Figure 4.8   The odd choice of colours on this cartogram makes a relatively simple distribution harder to grasp. Had some colours matched the countries national flags, blue for Scotland, red for England and perhaps yellow (for the daffodil) for Wales, the distribution would have looked far more obvious. Here areas are shaded white where high proportions are born overseas.

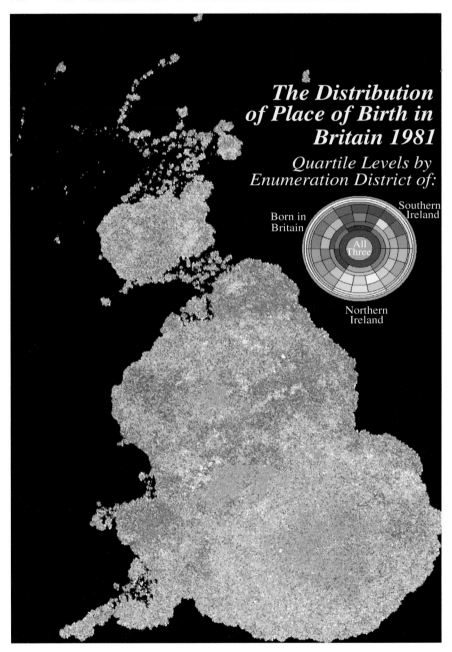

Figure 4.9   Here reds are used for places with high proportions of people born in Britain (above median levels), blues for people born in the Irish Republic and yellows for where high numbers were born in Northern Ireland. London, Birmingham, Coventry and Manchester show blue/yellow mixing (green). More people to the East of London were born in the Republic of Ireland than in the North of that island.

As the indigenous population was divided into three, so too can people born overseas be subdivided into three broad geographical categories (Figure 4.10). Here we use red for those born in Asia, blue for those born in Ireland and yellow for African and Caribbean born. This image is dominated by black areas, places with high proportions of all three overseas immigrant groups. Black represents the mixing on paper of all three colours, just as white represents the relative absence of people born overseas.

Tints, tinges and trends of colour in the image graphically show how the mixing varies.[7] The East side of the West Midlands is more Asian, the West more Irish in the backgrounds of its people. Those from Africa and the Caribbean settled in greater numbers in the South than the North of Britain, and so on. These are simple pictures; each block of streets (forming an enumeration district) is just a coloured dot, but already the combination of dots reveals, in a picture, details of the diversity of our society, which a search of the literature (see References) and conventional images failed to find.

4.4   Lost opportunities

Although the percentage officially unemployed in Greater London is a little smaller than average for Britain the city holds the largest concentration of unemployed in the industrialised world, and the real total is at least 150,000 larger than the total of over 400,000 admitted by the Government.

(Townsend with Corrigan and Kowarzik, 1987, p. 29)

Much of the most basic geographical nature of employment is largely determined by the first two questions addressed above: age and sex. Children do not (officially) work full-time and the elderly are usually retired. Traditionally, men get more paid work than women and what end up often called immigrant areas have often become (if they were not already) those places with the worst prospects of work. Therefore pattern builds upon pattern and we have to dissect the body of information we have with various statistical cuts and then rebuild it to a better understood whole.

The simple distribution of the proportion of the population unemployed shows strong connections with aspects of those distributions already mentioned above. 1980s employment and unemployment is another major theme that will run through this book. If we delve further we can compare the proportions

---

[7] The less aggregation the better: 'Further shortcomings exist in census data relating to ethnicity. Dissimilar birthplace groups are frequently aggregated into a single category: for example, all those born in the American New Commonwealth (chiefly the Caribbean) are usually grouped together in the published statistics. More seriously, several cross-tabulations in both 1971 and 1981 SAS group all New Commonwealth-born together. Prandy ... has demonstrated that the "social distance" between Asian and West Indian groups living in Britain can be as great as that between either of these groups and the British-born' (Ballard and Norris, 1983, p. 105).

Figure 4.10   A strong set of patterns is shown here. In 1981 half of Britain was practically empty of people who were not born on this island; the rest of the country presents some mixtures. London holds the majority of lifetime migrants of all groups and hence appears as a more colourful mass. The West Midlands shows a tendency for more Asians to live on its western side, Irish on its east.

of economically inactive people, mostly housewives, the retired, students[8] and children – sometimes somewhat prejudicially called *dependant* – with the unemployed and working populations, again using a three-colour scheme (Figure 4.11).

Red is used now for the unemployed, blue for those working and yellow for the dependants. These colour choices are important; they change the immediate impression gained from any image, if not any of the actual information in it. If there are not obvious colour matches, such as with England, Scotland and Wales, or with political parties, then it can help to match other variables on to some of those previous matches.

Areas where people are unlikely to find it hard to find work tend to more often vote Conservative, so matching high rates of 'employment' to blue colours makes some sense. Similarly, where work is scarce, 'unemployment' is high and Labour voting higher, so red is an obvious choice. Once two colours are assigned there are no more choices to make, the 'inactive' end up coloured yellow.

The picture of how the three labels of economic status combine shows much variation and a complex geographical pattern. Orange areas are those with high proportions of both unemployed and dependant people (such as in the Welsh valleys), green indicates many working and dependant people living in the same places (the Home Counties) and purple shows blocks where high numbers of people are working while many others are simultaneously unemployed (parts of London, for instance, where there are relatively few dependants).

Without some sophistication in colouring we might not have realised that often areas of simultaneously high employment and high unemployment could exist, how they coalesce geographically into clumps and how those clumps of many people who are working and unemployed, but with few people who are old or young or otherwise *dependent,* surround the Capital more than any other area of Britain.

Traditional maps entirely fail to portray distributions such as unemployment in any way that can be described as meaningful (Figure 4.12). They suggest a massive divide between the periphery and core by emphasising the fate of those living in rural areas – Scottish crofters as against London stockbrokers. In fact there are more unemployed people, and stronger concentrations of them, in the South than people living in the whole of the North East of England.[9]

---

[8] 'A third factor contributing to the large inflow into the South East is that students make up about 15 per cent of all immigrants and London is popular with overseas students as a place of study. The first and third of these factors go some way to explaining the larger than average outflow from the region. Outside the South East, the West Midlands and East Anglia were the most attractive areas for immigrants, relative to their populations. The relatively least attractive place for immigrants were the North of England and Northern Ireland (though it must be remembered that the figures take no account of immigrants from the Republic of Ireland)' (Davis and Walker, 1975, p. 5).

[9] Inequality is the crucial ingredient of deprivation: 'We shall hold that the most severe deprivation exists where the scores of disadvantage are high, where they affect the largest number of people, and where there is the most crass contrast between these areas and the advantaged periphery' (Begg and Eversley, 1986, p. 55).

Figure 4.11   These distributions show a well-defined set of patterns. Concentrations of both unemployed and inactive (orange) adults dominated many areas of the country including all the large inner cities and the Welsh coalfields. The South coast was highlighted by high proportions of the inactive (mostly retired) group. And in rings around London and other cities those in work dominated.

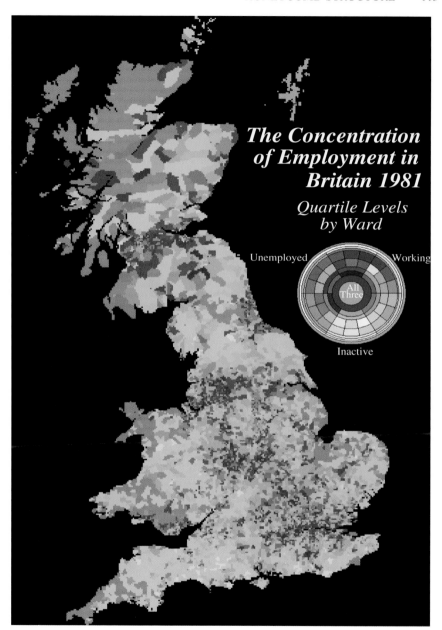

Figure 4.12   A very different image using the same statistics as the previous cartogram. Now the middle of England is dominated by the working and retired living in the same (green) places. The coast and country is where the inactive reside. Jobless people can be seen in the more remote areas, but the unemployment within the cities is eclipsed by the employed on their (bright blue) outskirts.

What often appear to be tiny unemployment concentrations on the equal land-area maps are great regions of joblessness in population space. The steepness of the slopes between nearby places of prosperity and places of poverty has always been of concern and a cause of fear for the more affluent. It is easier to forget more distant disparities. In 1980s Britain the depressed areas often abutted very close to the most fortunate.

When those who work (and did work) are subdivided into what they do for a living, there is not such close spatial affinity (there is remarkably little colour mixing in Figure 4.13). Again, we are limited to three categories to show how social classes do not mix, by the ability of our eyes and the flexibility of our imaginations, but three groups is enough to form a strong impression of the essence of social spatial structure. The divides in Britain are not so much between work and nonwork, but what kind of work is done by those in work.

4.5   Work, industry and home

The difference between inner Birmingham and the West Midlands southern fringe is... – the steepest in the country. Less than 10 miles separate some of the worst conditions in the country from some of the best.

(Begg and Eversley, 1986, p. 75)

Occupations can be divided into three broad groups according to how much people are paid.[10] These groups correspond closely with the general nature of the work (Box 4.2). The group commanding the highest income consist of professionals: managers, employers or landowners, often university educated. The middle section is termed intermediate in this book and includes foremen, technicians, skilled labourers and those in well-paid white-collar jobs. The lowest paid group are the supervised, made up of (cruelly labelled) unskilled workers, agricultural labourers and those poorly paid in basic service jobs.[11]

These three social classes of employees and the self-employed, namely professionals, intermediate and the supervised, correspond roughly to upper and middle class, lower-middle class plus upper working class and the remainder who do working class jobs. When blocks of streets are coloured by the proportion of

---

[10] The constitution and aggregation of classes is a contentious issue: 'The traditional distinctions between "manual" and "white-collar" labour, which are so thoughtlessly and widely used in the literature on this subject, represent echoes of a past situation which has virtually ceased to have any meaning in the modern world of work (p. 325)' (Braverman, 1974, quoted in Hamnett, 1986, p. 393).

[11] The grouping used here is similar to one used by Hamnett (1986): 'It makes little sense to aggregate such divergent groups and tendencies together and in the analyses which follow, SEGs 12 and 14 are treated separately from SEGs 8 and 9, on the grounds that they have more in common with SEGs 1, 2, 3, 4, 5 and 13 than they do with 8 and 9. Similarly, SEG 6 is analysed together with SEGs 7, 10 and 15 on the grounds of skill levels, remuneration and intercensal comparability. If this is not done, any comparison over time, let alone a sensible and meaningful comparison, is indeed virtually impossible.'

Box 4.2   Working definitions

The occupation groupings used in this book are defined by OPCS (1981, pp. 24–29). The three combinations chosen (using the New Earnings surveys of 1971 and 1981) were of socioeconomic groups:

Professional

1: Managers in central and local government.

2: Managers in industry and commerce.

3: Professional workers – self-employed.

4: Professional workers – employees.

13: Farmers – employers and managers.

Intermediate

5: Ancillary worker, artists, foremen and supervisors.

8: Foremen and supervisors – manual.

9: Skilled manual workers.

12: Own account workers (other than professional).

14: Farmers – own account.

Supervised

6: Junior nonmanual workers.

7: Personal service workers.

10: Semi-skilled manual workers.

11: Unskilled manual workers.

15: Agricultural workers.

The industrial groups were taken from the following amalgamations of 1980 Standard Industrial Classification based codes, referred to as 'Broad Industrial groups':

Code	Description	NOMIS class
1	Agriculture, forestry and fishing	0
2	Energy and water supply	1
3	Manufacturing industries	2–4
4	Construction	5
5	Distribution, hotels/catering, repairs	6
6	Transport/communication, banking, finance	7–8
7	Public administration and defence	91
8	Other service industries	92–98

Figure 4.13   The three colours mix little in this cartogram, but appear separately in every region and city - twisted around each other. The prosperous areas are coloured blue, the poor are coloured red. The middle of the country is dominated by skilled manual workers, often including miners, still relatively well off in 1981. This picture shows the local class structure throughout Britain.

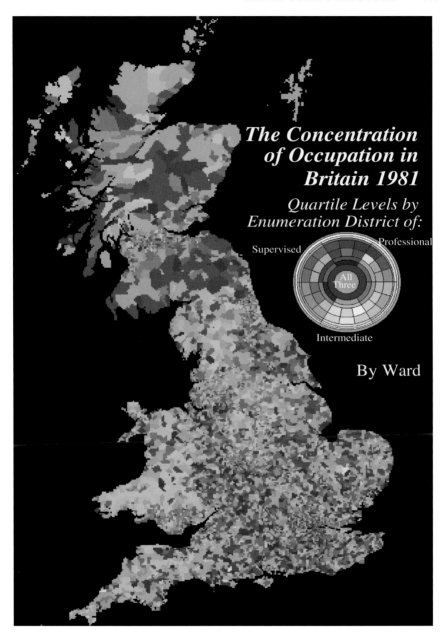

Figure 4.14   Again the same statistics are here shown on a map instead of a cartogram. Now we see a nation dominated by usually well-paid people working in professional and intermediate occupations. A few places, such as the remote areas of Scotland, are not doing too well. Only tiny areas of the map are allocated to the areas where the majority in fact live, but often earn much less.

the households whose heads belong to the categories blue, yellow and red, you can then see one of the most basic divisions of the social landscape[12] – the geography of class (see Figure 4.14). In this case it is obvious which colour goes with which class.

The colours do *not* mix. The blue *professional* band threads its way, lace-like, around the city suburbs; it is strongest in the South. The red *supervised* masses mark out the centres of major settlements, while the yellow *intermediate* patches show the distribution of relatively well-paid workers between the other two groups; in the coalfields of Wales and the North, for example (this map is drawn of a time before most of the miners lost their jobs). London is a city split between the most and least rewarded workers, with little room in between, a city of high and rising, but not yet extreme, inequality.

Smoothing the picture makes it easier to form some generalisations. Generalisation can be justified, in this case, because the information on occupation is only available from a sample of one-tenth of the population. Smoothing, when done evenly over population space on the cartograms, averages people with their nearest neighbours. However, the technique must be used sparingly if it is not to provide false conclusions. It should also be remembered that it is only the use of the population cartogram that allows the most poorly paid third of the population to appear in the picture other than as pinprick marks (Box 4.3).

Particular industries and those who are concentrated into urban areas to work in them, such as in high finance, can be pinprick sized on a conventional map. Graduates become attracted to certain areas and repelled from others; women and men find it more or less financially essential to work depending on what work there is in each area and what roles men and women are most expected to play at each time – in how they are expected to partake in industry. Like everything else, industry is geographically distributed, that distribution being important to the fundamental infrastructural geology of the social landscape.

Industry can be divided in many ways, for example into primary, secondary and tertiary, or into public and private sectors, and the voluntary remainder. Its distributions can then be painted. If this were done these images would now be the distributions of where people work, rather than where they live, a point addressed later in Chapter 6.

Instead of mixing three primary colours, a stratified geological type classification can be adopted, showing which industry has a majority of the workforce in each area. This industrial – geology colour schema has the advantage of further possible subdivision into dozens of industrial classifications, using subtle shades of the basic hues. It is also appropriate because one industry tends to be deposited upon another over time, but industry also becomes tilted, folded

---

[12] In the early 1980s class structure was not thought to differentiate cities much, although the following words were written before the 1981 Census results had been released: 'In only one of the largest cities (Liverpool) did the proportion of semi- and unskilled exceed the national average by more than 4 per cent. If concentrations of the most disadvantaged have occurred as a result of selective decentralization then it would appear to exist at a more localized level within cities' (Goddard, 1983, pp. 12–13).

Box 4.3 Two-dimensional smoothing

In some prints in this book the pixel maps have been smoothed by several passes of a binomial filter. In one dimension this can be written as $(1/4, 1/2, 1/4)$ and dissipates the intensity of a pixel with a value of 1 by the following intensities after the first five passes:

$$\left(\frac{1}{4}, \frac{1}{2}, \frac{1}{4}\right)$$

$$\left(\frac{1}{16}, \frac{1}{4}, \frac{3}{8}, \frac{1}{4}, \frac{1}{16}\right)$$

$$\left(\frac{1}{64}, \frac{3}{32}, \frac{15}{64}, \frac{5}{16}, \frac{15}{64}, \frac{3}{32}, \frac{1}{64}\right)$$

$$\left(\frac{1}{256}, \frac{1}{32}, \frac{7}{64}, \frac{23}{128}, \frac{35}{128}, \frac{23}{128}, \frac{7}{64}, \frac{1}{32}, \frac{1}{256}\right)$$

$$\left(\frac{1}{1024}, \frac{5}{512}, \frac{45}{1024}, \frac{5}{128}, \frac{95}{512}, \frac{29}{128}, \frac{95}{512}, \frac{5}{128}, \frac{45}{1024}, \frac{5}{512}, \frac{1}{1024}\right)$$

The two-dimensional version of this filter is given by the following matrix (after Tobler, 1969):

$$\left\{\begin{array}{ccc} \frac{1}{16}, & \frac{1}{8}, & \frac{1}{16} \\ \frac{1}{8}, & \frac{1}{4}, & \frac{1}{8} \\ \frac{1}{16}, & \frac{1}{8}, & \frac{1}{16} \end{array}\right\}$$

After approximately ten passes this filter is equivalent to the effect of a normal kernel with variance $n/4$ (where n is the number of passes). This is one of the simplest and most elegant forms of spatial smoothing. It is also, interestingly, reversible (although this is only practical for low numbers of passes). Its inverse could theoretically be used to sharpen an image.

and faulted. The pictures produced show the clear divisions between the sectors and strong geographical patterns. These underlie many of the images already presented and many of those still to be seen.

It may appear odd today to see maps in which transport and banking are put within the same category. However, in the 1980s they were just seen as different parts of the general business and domestic support infrastructure, mundane, run-of-the-mill, industries. Later on banking came to be categorised separately and seen as an activity that could somehow make money in its own right, not simply by providing some basic services. This had many great adverse consequences.

A further consideration here is that the kind of work people do affects the sort of home they will have. Price can indicate the quality of a house, the social status of the neighbours, as well as, perhaps, the inflated and depressed states of local markets. During the 1980s it became more obvious that price was more about place than the quality of bricks and mortar, and housing price speculation became more closely associated with gambling. This was most evident when housing prices crashed towards the end of 1989.

It must be remembered that in examining housing, what are being considered in terms of price are the prices only for homes that are for sale. The prices shown here thus illustrate the distribution of privately owned homes being bought with loans.[13] Average housing price has been estimated and plotted for wards from a sample of building societies' records (Figure 4.15). The geographic patterns of local housing price structures are investigated later in Chapters 5 and 8. For now, the close correspondence, and important differences, between this picture and the others are all presented together to illustrate the closeness of the connections.[14]

4.6   How people vote

> *You can no more take politics out of government than you can keep*
> *sex out of procreation.*
>
> *(Gyford, Leach and Game, 1989, p. 1)*

The British state regularly asks its inhabitants for their opinion on its government, through elections of candidates representing political parties standing for particular issues. As the choice is usually only between two or three regular parties this expression of choice is extremely limited. During the 1980s most people's choices were, respectively, the Conservative, Labour and Liberal parties, which respectively adopted the colours blue, red and yellow to represent themselves.

In the early 1980s the Liberal party allied with a newly created Social Democratic party. Following 1990, mostly merged, they took the label Liberal Democrats. The term Liberal is used wherever possible in this work to avoid confusion. Other parties, such as the Nationalists, Unionists and Greens, showed

---

[13] 'Social class, in the sense of status of individuals in the labour market, may today be as well reflected by position in the housing market as by necessarily imprecise occupational labels' (*Buck et al.*, 1986, p. 101).

[14] The distribution of housing is intricately connected to many of the other patterns shown here: 'The results of this study strongly support the argument (Cheshire, 1979) that inner-city unemployment is not so much a problem of the performance of the city labour market as a whole, but a feature of the other sifting mechanisms in society, mainly the housing market, that concentrate people who are at a competitive disadvantage in society into relatively restricted areas' (Frost and Spence, 1981, p. 100).

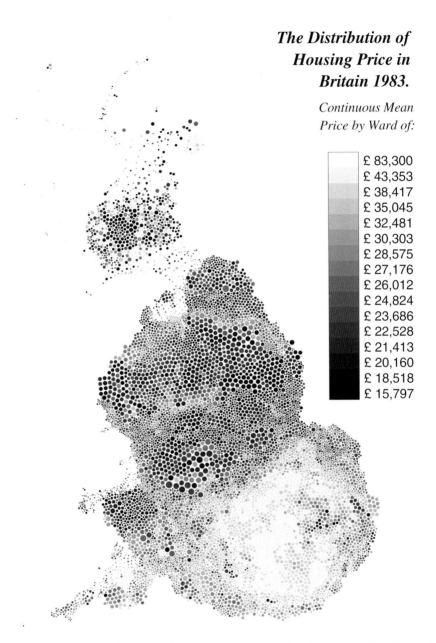

The Distribution of Housing Price in Britain 1983.

Continuous Mean Price by Ward of:

	£ 83,300
	£ 43,353
	£ 38,417
	£ 35,045
	£ 32,481
	£ 30,303
	£ 28,575
	£ 27,176
	£ 26,012
	£ 24,824
	£ 23,686
	£ 22,528
	£ 21,413
	£ 20,160
	£ 18,518
	£ 15,797

Figure 4.15   The paler area of more expensive housing in Greater London and the Home Counties is extremely distinctive, but at this spatial scale a few wards of very low local housing price can still be seen, clustered in the East End. Glasgow city is also particularly distinctive. Prices over a number of years would produce a more reliable but similar overall impression.

interesting, but not particularly influential, geographical distributions during this period.[15]

Study of the human geography of voting is well suited to equal population visualization (Figures 4.16 and 4.17). Cartograms give people the equal representation their votes are worth, and the graduated three-colour scheme encompasses most eventualities. Although only the winning candidate holds each seat in parliament, the degree of support they command and nature of the opposition to them are also relevant both for one point in time and to indicate possible future trends (Figures 4.18 and 4.19).

The votes in national (general) elections are only reported for very large areas containing sixty or seventy thousand electors. While general elections are pertinent events, and the only complete record of the people's (or at least of those who do vote – Figure 4.20) actual wishes for government, the fine detail of local opinion at which we know everything else about our social landscape is lost when voting preferences are reported only at constituency level.[16] Voting statistics for Westminster parliaments are not released below the constituency level.

Local elections follow a complicated system of timing and are not all simple one-candidate outcomes. They do, however, give us information, at the relatively fine level of several thousands of local contests, nationally and annually. County council and Scottish regional elections, however, are based upon a different, very poorly defined, geography ('County Divisions'). Therefore we must rely upon the results of District elections to see the spatial mosaic.

The national picture of voting emphasises the divisions seen earlier in the social landscape. In general people in areas of high unemployment, recent immigration and older industries vote Labour, while the rest of the country is dominated by the Conservative party, closely followed by the Liberals (Figure 4.21). The geography of political party support accentuates the differences between people grouped by area, as it has traditionally been through these parties that people are allowed to register their support or condemnation of the social system within which they are placed.

Cartograms much simpler than those shown here are just beginning to be used by the media. On television pundits appear to touch a screen and zoom in on areas already given equal weight on election night TV population-based maps.

---

[15] Very small parties can have very big effects, even under the first-past-the-post system as political bias can be in either direction: 'The January 1910 election illustrates this situation very clearly. The Irish Nationalists won 82 seats, all but one being located in Ireland. ... Thus with 1.9 % of the vote the Irish Nationalists were able to secure 12.2 % of the seats to enjoy the positive bias of 10.3 %. The more recent experience of the Liberals has been a sharp contrast to this situation' (Gudgin and Taylor, 1973, p. 18).

[16] The local distribution of class was once seen as almost identical to that of local voting: 'Since 1945, occupational class has been widely seen as the main social basis underlying electoral politics in Britain. A pattern of "class alignment" was clearly apparent in the 1950s and 1960s. ... at the level of explaining why particular areas or constituencies vote the way they do, knowing the mix of occupational classes in the local area continues to be as valuable as ever in explaining or predicting election results' (Dunleavy, 1983, pp. 32, 37–38).

Newspapers put up interactive cartograms on their websites to show the results of those same elections and to try to lure in more readers. The next step will be to see the same kinds of cartogram being used to map less obvious measures than votes.

4.7   The social landscape

The Victorians were more concerned with patterns found at a smaller spatial scale, and this is also reviving today: as we will see, despite the interest in the north–south divide, some writers see spatial differentiation taking place on a much smaller spatial scale between localities.

(Savage, 1989, p. 248)

A particular representation of the social landscape of 1980s Britain has been built up and presented through a dozen colour images. This is the landscape that is made up of, and to varying extents determines, many aspects of peoples' lives. It is the landscape of neighbourhoods, communities, blocks, streets, groups, villages, suburbs, housing estates, life chances and constraints. It is the landscape of age, work, class, immigration and race. It is the landscape of social existence, political power and economic opportunity – the human geography of Britain.

There is, however, much more that could be studied using these methods than has been shown so far, all adding to the montage of a social landscape. The geography of health is one area – of people's life expectancies, of disease and disability; the geography of welfare – the payment of benefits, the provision of services; the geography of privilege – the distribution of power in the workplace; the ownership of property; shares in industry; the geography of income and wealth.[17]

We know where most ill-health is to be found (among the old and poor), where the most social security benefits would be paid, where the rich would be found and where the owners of industry concentrated. As you draw more and more of these pictures, you begin to recognise the same familiar features over and over again in the social landscape. So much is so strongly inter-related that

---

[17] Some of the patterns found here were previously gleaned from more conventional maps of Local Labour Market Areas (LLMAs): 'Most notable is the ring of most privileged LLMAs around London, extending to the South Coast and forming a virtually complete arc on the other flank; the only exception being along both sides of the Thames estuary. There are also significant clusters of better-off LLMAs further westwards along the South Coast and in southern parts of the West Midlands. The prosperity of the South Coast can be gauged in terms of the fact that south of a line between the Severn Estuary and Lincolnshire there are only three representatives of the lowest quintile on this indicator, namely Corby, Spalding and Wisbech, and, of Britain's bottom 112 places, the South accounts for only 11, all but one of which are located on the margins of the region in East Anglia and the East Midlands (Figure 7.1A)' (Champion *et al.*, 1987, pp. 91–93).

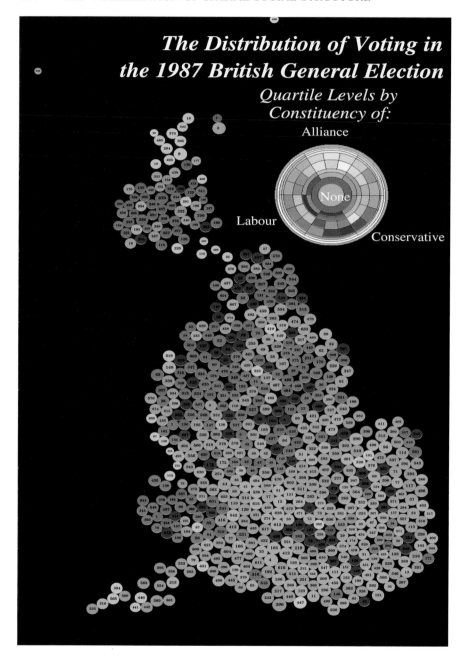

Figure 4.16   This image is based upon quartiles in a three-colour circle, rather than the more appropriate electoral triangle, which is used later. The distribution revealed resembles that of occupation, but London is a slightly lighter shade of pink than you might expect (a weaker Labour vote than the job profile of London alone explains). Most blue seats are in the South.

Figure 4.17   The map contains the (occasionally obscured) full names of constituencies. It shows a very different picture to the cartogram. England is covered by the blues, greens and purples that spell out a Conservative landslide, while the yellow and lime greens of Scotland and Wales show us Liberal country. The red of the Labour party is relegated to a few cities and the Welsh Valleys.

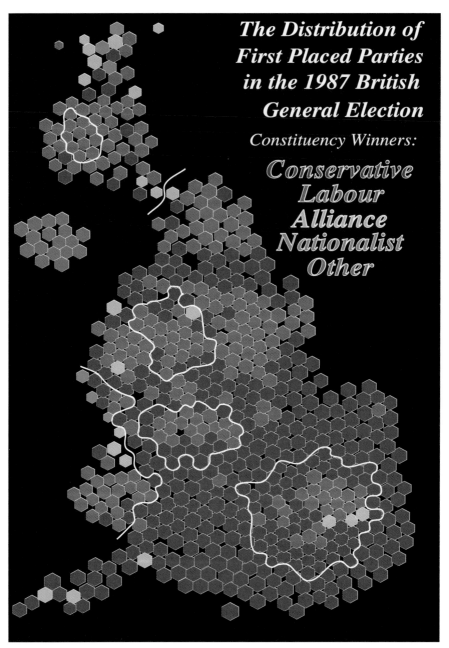

Figure 4.18   Showing all 650 UK parliamentary constituencies, the boundaries of the major conurbations and nations are included. A few constituencies are isolated from the main groups – Oxford East and Norwich South for Labour, Tynemouth and Edinburgh West for the Conservatives. The Celtic fringe of Liberals, Nationalists and other parties in Northern Ireland is easily seen.

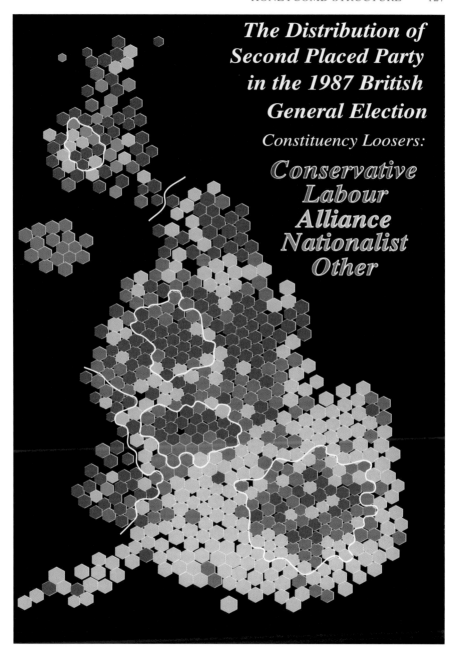

Figure 4.19   This is a much more interesting picture than the common one of victory. The ring of Labour misses, between a blue central and orange outer London is (in hindsight) telling, as is the similar pattern around Birmingham. Nationalists came second more often in Scotland than in Wales. The picture looks behind the simple message of electoral success towards who might win in the future.

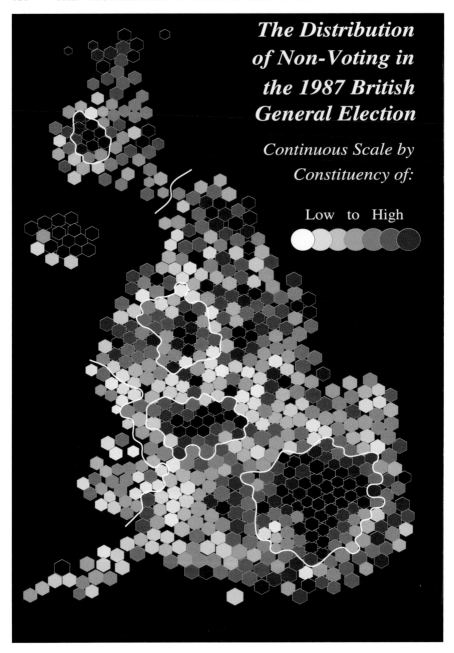

Figure 4.20   Another picture rarely seen. The major cities and most of Northern Ireland have low turnouts, but so do the (elderly) south coast and (remote) Scottish highlands. High turnout can be seen in some of the Welsh Valleys as well as in more prosperous areas. Marginal seats had a higher turnout than those that were safer, but this image does not simultaneously show marginality.

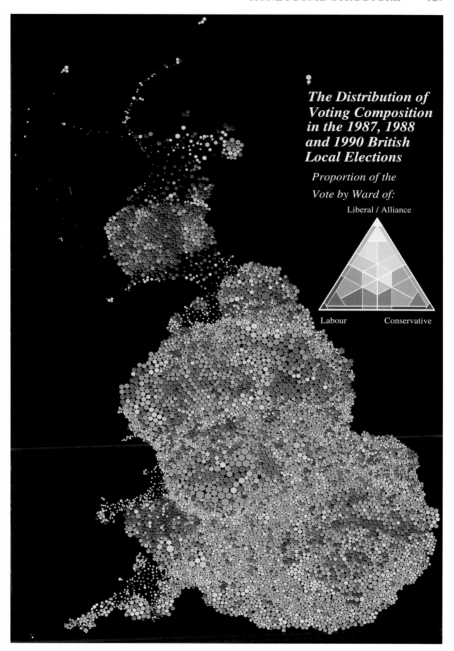

Figure 4.21   The triangle shows the colours for the safest seats at its corners, easy wins next to them, two-way marginals in the middle of the sides and three-way marginals at its centre.The cartogram clearly shows these results for all 10 444 wards, small and large. The pattern shown is similar to that of occupation. Local party strongholds can be identified.

it becomes the exceptions – which point to how things can be different – that appear most worth looking for.[18]

So much for the static image. The next question to ask is: now that we can see the landscape, how is it changing? Then, what movement keeps it alive and what alters it? These are questions that stretch the limits of visualization as a methodology. In the final part of this work the problem of envisioning the history and geography of social structure simultaneously will be addressed, suggesting possibilities for new methods for the integration of all this information to create a consolidated image of British social history and geography.

Time and space intermingle. The pictures presented up to now show the situation at a single point in time, but those of immigration and birthplace (just shown) also tell of a different past, aspects of change that are the focus of the next chapter.

---

[18] 'This, then, is the South–South divide. It is a divide which appears in employment opportunities, in wage packets, in each job's content and potentialities. It reappears in the car park and the bus queue, in the green of the garden and the size of the room. Each part of the divide has its own daily timetable and its own life cycle' (SEEDS, 1987, p. 10).

5

Transforming the mosaic

Yet within any town as in the region as a whole there is a pattern. The poor housing, schools and levels of unemployment will tend to be concentrated in certain districts – as they are concentrated in inner city areas of the large conurbations of this country. At the level of the region, too, there is a pattern, increasingly clear and changing.

(SEEDS, 1987, p. 6)

5.1 Still images of change

Counts, measures, votes and all the other figures we use to build a picture of our social landscape are collected regularly because it is recognised that the picture changes.[1] People's positions, locations and aspirations alter. Compared to the static picture, much less research has been addressed to looking at change. Areas are classified as being rich or poor, but hardly ever as becoming richer or poorer. It might be said that they are gentrified, or are now a slum, but not that that is the direction in which they are heading.

Areas being classified by what they are, rather than what they are becoming, may be partly the result of the change generally being slow and also very uncertain, but much of the reason is the difficulties of displaying change. These difficulties range from simple problems of gathering the information (temporal

---

[1] 'Our maps are in one sense diagrams of geographic systems and their evolution. Many of them are – or were – cartographically communicated theories about global or regional geographic systems of resources and settlement. Many time series of maps are in one sense statements of theories, in cartographic language, about geographic development processes, about the functioning and the past and future evolution of some global or regional geographic system' (Borchert, 1987, p. 388).

---

discontinuities), to differentiate sampling error from change, to the technical difficulties of processing it (what type of change is to be seen), to, finally, the imaginative hurdles that have to be crossed in portraying it (creating still images of change).

5.2   Forming the structure

The decennial census ... was a solution to the problems of data collection and dissemination of the mid-19th century. Yet we have only the vaguest notions of how to exploit the new technology in this area; our perceptual systems are so geared to conventional display ....

(Goodchild, 1988, p. 318)

In past research, geographers often stumbled at the first hurdle when gathering information about change; this hurdle was coping with temporal discontinuities. Temporal discontinuities occur when units of population have their spatial boundaries altered. Temporal discontinuity is continuously reoccurring, and is itself one aspect of change. As people move, so do the collating boundaries move around them, eternally attempting to encompass them adequately. We need to encompass these changes within our pictures.

Practically every person in Britain is counted in the census every ten years.[2] The simplest single number to be gathered from this is the total change in population, an increase of however many thousands between 1971 and 1981. How is this simple loss or gain of people distributed over our landscape? During this period Britain undertook its greatest ever redistribution of administrative boundaries – everything altered. Very few figures collected before the mid-1970s could be directly compared with those that came after. The changes in 1965 and 1974 were as great as all those from 1975 to 2011.

Geographers often addressed the problem of change with the crude solution of aggregation.[3] The method is to find a set of large areas either side of the time period whose summed figures can be directly compared. This solution causes a

---

[2] Many minor nuances must be included when calculating the change between censuses: 'Perhaps most fundamentally, the 1981 Census was taken on the night of 5 April (before Easter and out of term for higher education institutions) and the 1971 Census was taken on 25 April (after Easter and in term for higher educational institutions). In towns where the number of holiday-makers and students cause seasonal fluctuations in the size of the population, this three-week difference is likely to have some impact on the results obtained' (Norris and Mounsey, 1983, p. 276).

[3] A policy of aggregating areas can prevent a proper study of geographical change: 'The largest tract in the region, in terms of 1971 and 1981 EDs, occurred in the district of Bracknell in Berkshire, with 98 1981 EDs and 60 1971 EDs combining to form this comparable "small" area (as defined by OPCS). This tract is therefore a good example of an area in which great change is taking place, but which – as a consequence – permits the least local study in the region of this change, due to the large size of the tract' (McKee, 1989, p. 4).

great deal of information to be lost; local patterns can no longer be seen and large areas arbitrarily appear uniformly good or bad when the more truthful picture is very different (Figure 5.1). When aggregation is employed it is often advantageous if the larger areas are home to similar numbers of people.

To see national patterns or regional or city-sized processes it is sometimes better not to use national, regional or city-sized spatial units. Rather, show the eye the finely detailed picture and let the mind decide how much pattern does or does not exist. Only then can the decision be made whether to smooth the picture further. Only if you've seen the detail do you know if the aggregation provides a fair summary. There are also many means other than indiscriminate geographical amalgamation that can be used to generalise an image.

How, though, do we create these fine images of local change from two sources based upon small, but differing, areas? One solution is to recognise that any change in boundaries has only a very local effect (Box 5.1). People are moved from one side of the line to the other. There is no need to abolish the line, simply to realise that a few people have been moved. A detailed image, where nothing but the boundaries has really changed, will simply appear a constant, slightly speckled shade. The eye interprets the fine dithering that will have been created by misplacement as a colour, not a pattern. The problem has been reduced away.

It is not always possible to say there is a simple solution to depicting change when you try to give all people equal representation. Between 1971 and 1981

Box 5.1   Linking the censuses

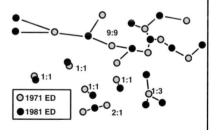

The 1971 and 1981 census geographies were linked at the enumeration district (ED) level. The majority of ED boundaries had not changed or were nearly identical, but in some places substantial alterations had occurred, for instance where a new town had been built or an old estate pulled down. The use of the 'census tracts' designed by OPCS had been found to be far from adequate by McKee (1989). An alternative, far more flexible solution was devised. Only enumeration district centroids were known for each set of roughly 130 000 points. Two two-dimensional tree-data structures were built and the closest 1981 district to each 1971 found, and vice versa. Thus every ED in each set was connected to at least one in the other, but could be connected to any number, if necessary. Every ED count could then be compared between the two censuses.

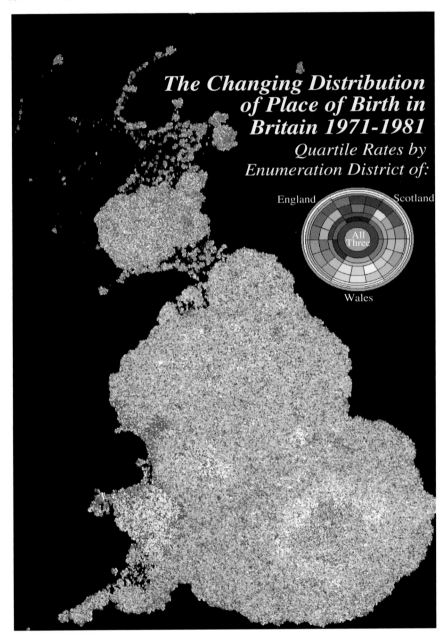

Figure 5.1   The most distinctive feature of this image is of the relative increase of the Welsh in the Valleys. South Wales had the lowest emigration rates and so appears to have increased its Welsh population. The whiter rings around London, Birmingham and inside Leicester are where the overseas born population has increased. Parts of Glasgow became even more Scottish.

we were fortunate that the total population, and its internal distribution, changed little. Advantage can be taken of this fact. For instance, the underlying cartogram that was used to portray the people of Britain did not, except in a few parts, perceptibly change its shape.

Over long periods, which are examined later in this chapter, the underlying population basemap changes with time. Over a much longer time period the underlying base of geological maps change too as coastlines shift and mountains rise. Just as over geological lengths of time usually a single landmass splits into two, or two are joined, so too in human geography, when looking at change over decades rather than millennia, most spatial units, of any size, will usually correspond to just one or two predecessors. Occasionally the relationships will be more complicated (Box 5.2).

Box 5.2   How closely connected?

The following tables show how often a one-to-one link was achieved and how it is unnecessary to combine numerous districts when this approach is taken.

Links between 1971 and 1981 Enumeration Districts

1971 EDs to each in 1981		1981 EDs to each in 1971	
0	0	0	216
1	105 946	1	100 436
2	20 550	2	20 841
3	2347	3	3 085
4	287	4	582
5	49	5	163
6	20	6	63
7	4	7	34
8	5	8	18
9	0	9	12
10	1	10	9
11	2	11–18	17
Total 1981 EDs 129 211		Total 1971 EDs 125 476	

Note: 216 of the 1971 enumeration districts had identical grid references of 600 km East by 400 km North, a place in the North Sea, and could not be linked to the 1981 set. These had been used to record the population off shore for those districts with a coastline or inland water.

5.3   Structure transformed

> *The inadequately described have moved almost exclusively into the council sector.*
>
> *(Hamnett, 1987, p. 548)*[4]

The most basic changes of population have been painted here simply by making each block white where population fell and black where it increased.[5] The white holes of the major conurbations can be easily distinguished, as can the black rings of built-up areas around them.[6] De-urbanisation was uncovered as taking place in 1970s Britain when the 1971 and 1981 censuses were compared. Importantly, the truth of this generalisation can be ascertained from just how clearly this pattern stands out in the images shown here.

There are no woodlands and fields on the population cartogram, just the 'people-lands' of inner cities, suburbs, small towns and villages – all drawn in proportion to their populations. Some of these have been growing and some declining. Everywhere there has been great variation, from street to street, suburb to suburb.[7] As the image is progressively smoothed, averaging each cell of one hundred people with eight hundred of their neighbours, then smoothed again to average over two thousand, then five thousand, then eight thousand ... a more and more generalized image of the process of population redistribution is revealed. This information is perhaps more clear, but less real.

The changing distributions of the sexes and ages can also be depicted with shading that is then smoothed. For the distribution of the sexes within any particular place to alter, people must be born, die or move. Age obviously changes continuously with time, as well as irregularly over space (as people move). These two attributes are, however, interrelated, for as people age, men die earlier and

---

[4] Textual description of change can also be very elusive as a result of generalisation. There was a category of people referred to as having their occupations 'inadequately described' in the censuses of 1971 and 1981 and, by 1981, this group had both grown in number and a large majority of the group were residing in council owned housing. The sentence quoted above sums up how people get lumped into intersecting social boxes that often says as much about the designers of the boxes, boxes both made of bricks and mortar and on paper census forms, as it says about the people who were boxed.

[5] When there was a population change, ward boundaries were rapidly redrawn: 'The most extreme examples are a new ward in the Isle of Dogs with a zero population in 1971 but 5,400 in 1981; and a ward in Bracknell district with a population of 3 in 1971 but 8,700 in 1981' (Craig, 1988, p. 9).

[6] It should be possible to see the regional pattern through the local picture: 'With the exception of the South East, in all regions containing a metropolitan county the balance of migration both in 1971 and in 1981 was outward; and in all the remaining regions it was inward' (Brant, 1984, p. 28).

[7] 'The City of London was the only London borough to increase in population during the 1970s yet it is precisely this district in which a number of tracts experienced some of the greatest decreases in population in the region during this period' (McKee, 1989, p. 201). The overall increase was due to the populating of the Barbican estate: http://en.wikipedia.org/wiki/Barbican_Estate.

so changes in the proportions of the elderly are reflected by changes in the geography of the sexes.

More children will be born and brought up where there are more women. Again one influences the other. We could struggle to see these influences on three separate maps, one of the elderly, one of women and one of children. But how much better is it to show these interrelated changes in a single image by three-colour shading? Although more confusing initially, with a little study and patience trivariate mapping reveals patterns that three separate maps cannot (you cannot merge them in your mind). Now, though, rates of change rather than proportions of people's movements and patterns of deaths and births are under scrutiny. What aspects of change should be highlighted? (Box 5.3).

Aspects of the changes discussed above can themselves be examined more closely. Change in the spatial distribution of children of different age groups was examined, but produced an image considered not worth printing here. This image, which presents such a jumbled picture, tells us that there has been little uniform progression in these five-year age bands over time. The confusion is caused by families moving; in many places only a minority remain in the same block for ten years.

Box 5.3   Measuring the changes

Observed change (O) can be measured in many ways between two times (T) and many places (i), for instance

$$O_i = \frac{T_{i2} - T_{i1}}{T_{i1}} \quad \text{or} \quad \frac{T_2 - T_1}{(T_1 + T_2)/2}$$

Expected change can be calculated by

$$E = \frac{1}{n} \sum_{i=1}^{n} \frac{T_{i2} - T_{i1}}{T_{i1}}$$

Then deviation from the expected is given by

$$D_i = \frac{E - O_i}{E}$$

Given six categories of housing and the national average prices (P) and proportions (W) of these, it is possible to calculate from the local distribution of average prices (p) an average house price (h) as either an arithmetic or geometric mean:

$$h = P \sum_{i=1}^{6} \frac{W_i \, p_i}{P_i}, \quad h = e^{\frac{1}{P} \sum_{i=1}^{6} \ln(\frac{W_i \, p_i}{P_i})}$$

Unravelling the effects of migration is difficult. However, 'no obvious pattern' can be an important image to show, especially when a pattern is thought to exist. The next chapter examines it in detail. For now, it is worth noting that even pictures that show no structure are showing something. Until you look at a picture you can only guess what you may or may not be able to see. However, if you look and see no pattern, you need not include the image, even if the fact of no pattern is interesting.

The last chapter included a short introduction to the major theme of migration by looking at where people in Britain were born. Now we look at how those pictures are changing (Figure 5.2). This could be done by seeing where those migrants have moved to, or from, but here we show how the proportion of lifetime migrants has altered in areas over the ten years between censuses. High levels of colour in the image indicate either that the proportion of lifetime migrants in that area has risen or that it has not declined as much as elsewhere.

Of those born in Britain, what is most striking is the return of the Welsh to the Valleys or, much more likely, their relative unwillingness to leave them. Less obviously, the image highlights an infusion of English born into the rest of Wales and into highland Scotland. The decline of all three national birthplace groups generally occurs in areas where people originally born overseas have been moving in or moving to.

The picture of change in the proportions of those born in (all of) Ireland, Asia and Africa depicts some interesting features (Figure 5.3). The rings of movement of lifetime international migrants out of the centre of London are distinctive.[8] This group of people had a major impact upon the changing social landscape of Britain in the 1970s as their moving out influenced many other distributions and made space that other overseas migrants were both attracted to and encouraged into.

5.4   Variable employment

Analysis of trends over the 1980s points to a continuation of wide differences between the least and most privileged wards. Unemployment differentials have widened, even in the most recent period when the average level has fallen.

(Congdon, 1989, pp. 489–490)

One of the most variable attributes of our social landscape is employment, or the lack of it. The obvious extension of the above methods is to divide

---

[8] Before the 1981 census results were released, and long before any questions were asked on ethnicity, which first occurred in the 1991 census, it was assumed polarisation would occur: 'The Black population of Britain is locked into an allocative system that seems bound to produce an increased polarization of native and immigrant populations. The forces that drew them into the economy are the same forces that are producing an increased isolation of the Black population. They came to fill gaps created by an upward mobility of the White population in the employment structure and they settled in gaps left in the urban structure by the outward geographical mobility of the White population' (Peach, 1982, p. 40). The later censuses showed this not to be occurring.

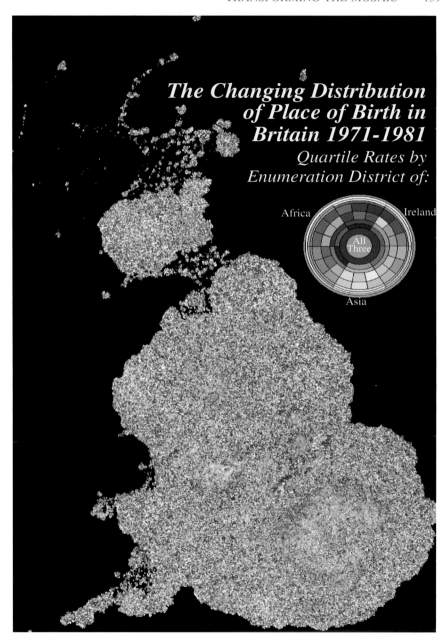

Figure 5.2   The orange tinted ring around central London shows where those of African, Caribbean and Asian origins have moved to. The whiter centre shows the places they were leaving. The yellow patches point to the influx of Asian (mostly Bangladeshi) immigrants in the 1970s to East London, to Sandwell in the West Midlands and to a few isolated parts of the North of England.

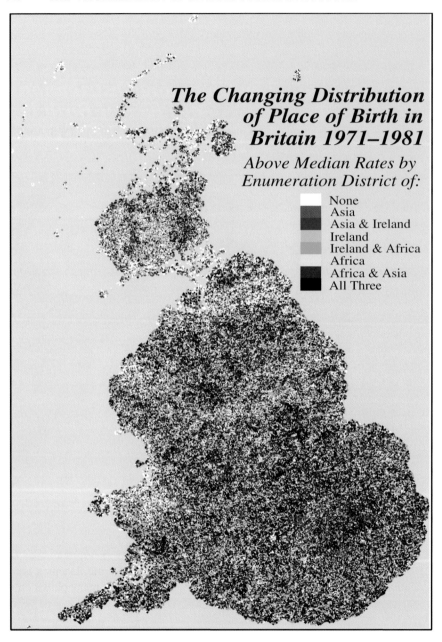

Figure 5.3   When the much simpler division of change into above and below average is shown a simpler set of patterns is revealed. There is much less immigration of people born outside of Britain into the North East of England. Of those who do go to the North, Wales or Scotland, in general they are more likely to have been born in Ireland. It is the other groups that mostly went to the South East.

the population into three groups: unemployed, working and the particularly interesting (and often ignored) remainder, and then show how (in each very small area) each group changed during the 1970s. This has been done here to illustrate how people's lives would be affected by changing levels of employment from 1971 to 1981, which depended greatly on where they lived (Figure 5.4).

The West Midlands was the largest area to have solidly increased its share of the unemployed and inactive during the 1970s; more people were working around London, but within the heart of London more of those deemed economically inactive were left, perhaps as households containing more members in work more often moved out to the suburbs.[9] These images show strong patterns, but – just as indistinct images are not worthless – simple ones are not necessarily true.

Employment is a feature of our social landscape that changes seasonally. If you like physical analogies it is the vegetation cover. The single change over ten years hides great swings in the fortunes of places between those dates. Unemployment has been measured for areas the size of towns for every month since 1978, providing what initially appears to be a very hard to understand spacetime series in comparison with simple 1971–1981 decadal change (Figure 5.5).

To show changes most simply we can paint a small image for each year, showing the deviation in each area from expected levels for that place and time. Such a series shows us how the spacetime pattern of unemployment deviates from what we would imagine, given a simple graph of time and a single cartogram of space (Figure 5.6).

The series of cartograms showing changing unemployment levels between 1978 and 1990 has areas shaded dark to indicate higher than expected levels of unemployment, moving towards white for lower than expected levels. At the end of the 1970s a Celtic fringe of high unemployment is apparent; by the end of the 1980s a very distinctive ring of low unemployment has grown around London.

The shading of the areas in this unemployment map is as dependent on the limits of the time period as it is on the spatial limits of Britain. What is more, only a few years can be shown on a page, although at least years are a sensible amalgamation of months for counting unemployment. What these images illustrate, from the point of view of visualizing social structure, is the beginnings of a combined picture in space and time of evolving unemployment rates.

The pattern (geology) of industry changes much more slowly than that of employment, even though the latter follows changes in the former. Detailed non-population-census information on people working in industry in many places only became available towards the end of the 1980s and only then for the years that the census of employment had then covered. The change for wards was shown earlier, as measured between the 1984 and 1987 employment censuses, and an

---

[9] It is easy to forget that the early 1980s were still a period of economic decline for London, although people more often moved out than signed on and stayed put: 'The long-term decline in employment in London – which goes against the national trend – and the even sharper decline in inner areas has come about without a significant upward shift in London's unemployment' (Buck, 1986, p. 180).

Figure 5.4   The term inactive is used by the census authorities to include those who are retired, permanently sick, students and housewives. Increasing unemployment causes more early retirement, less employment of the disabled who then more often register sick, other measures used to keep the headline unemployment figures down and, more now than then, more staying on in education.

alternative colour scheme is used here to depict which industries grew the most and which showed most decline.

The distribution of occupations, as might be expected, hardly altered at all over the ten years to 1981 (Figure 5.7). Therefore, although employment rates can rise and fall quickly, industries come and go more slowly. Those areas that tend to house the managers who are in charge, and the workers they are in charge of, tend to remain much the same for far longer, even as what is being done alters from bashing metal to pushing paper.

5.5   House price inflation

> ... if 'economics is all about how people make choices, sociology is all about why they don't have any choices to make' .... For many, the idea of a free choice in housing is a sick joke, especially amongst the unemployed, those in insecure jobs, and for many in high house-price areas.
>
> *(Murphy, 1989, p. 101)*[10]

Before the 1980s we could not draw detailed national maps of house prices changing across many small areas; we simply did not have good enough data, let alone a way of mapping it. For the maps produced here a large sample of house sale information has been collected for the years 1983 to 1989 inclusive. This has been converted for processing at the ward level by linking every postcode used then to a grid reference and hence a ward.

The static picture of housing price has been shown before in this book; now inflation at every place in every year is shown. Here, a different shading scale from that used above is adopted. Understanding housing price change involves comparing years as well as areas, so fixed continuous shading is employed: light to indicate rapidly increasing prices, dark for falling ones.

The picture is at first murky, a problem, perhaps, with the somewhat unreliable figures typed into the dataset by building society clerks in the first year mapped (Figure 5.8). With so few homes and such different houses and flats being sold by particular building societies in particular wards in particular years, spurious price changes can be suggested. To produce the maps shown here the mix of houses being sold in a ward is weighted to match the national mix before a weighted mean housing price is calculated for the area.

---

[10] Here Murphy is quoting Rex and Moore (1967), Duesenberry (1960) and others. In full the quotation reads: 'A rational choice model is clearly oversimplified: the idea that couples have a free choice between sectors is – in both senses of the word – untenable: access to different types of housing is determined by "constraint" as well as "choice" (Rex and Moore, 1967). In Duesenberry's words if "economics is all about how people make choices, sociology is all about why they don't have any choices to make" (1960, p. 233). For many, the idea of a free choice in housing is a sick joke, especially amongst the unemployed, those in insecure jobs, and for many in high house-price areas' (Murphy, 1989, p. 101). The 1960 and 1967 references are given in the original source.

The Space/Time Trend of Unemployment in Britain, 1978–1990.

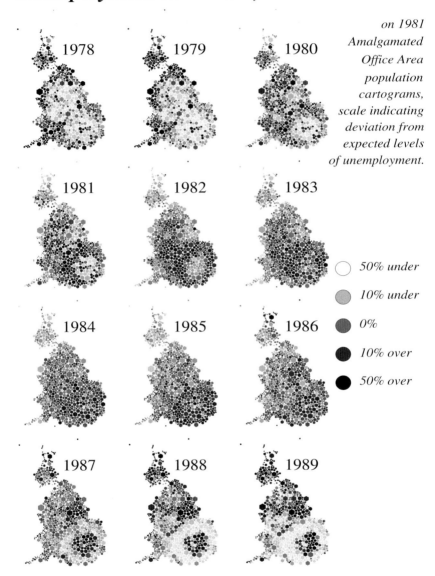

on 1981
Amalgamated
Office Area
population
cartograms,
scale indicating
deviation from
expected levels
of unemployment.

○ 50% under
◔ 10% under
● 0%
● 10% over
● 50% over

Figure 5.5   Twelve years of unemployment figures, shown using a cartogram of the 852 amalgamated office areas. A very distinctive pattern emerges of 'Celtic Fringe' turning into 'Inner City' decline against a backdrop of a ring of relatively good fortune was forming around the Capital. The pattern is well defined and the area getting worse most quickly is seen then to be central London.

The Space/Time Trend of
Unemployment in Britain, 1978–1990.

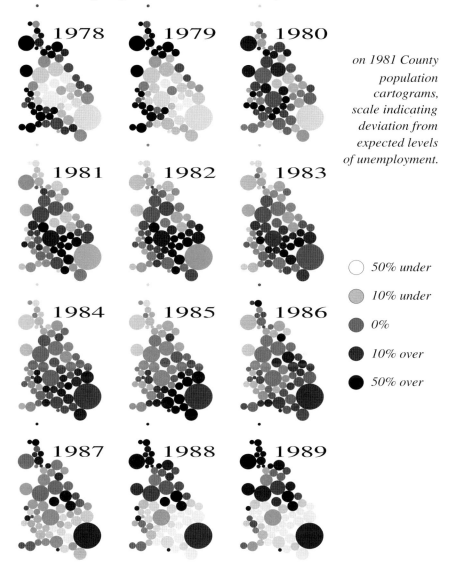

*on 1981 County
population
cartograms,
scale indicating
deviation from
expected levels
of unemployment.*

○ 50% under
◔ 10% under
● 0%
● 10% over
● 50% over

Figure 5.6   It is easier to compare a series of cartograms when there are fewer areas – 64 counties here – and some with distinctive sizes. Yet part of the impression here is not correct but an artefact of the areal units chosen. The black spots in London are partially cancelled out by nearby very prosperous areas. The southernmost spot, dark in later years, is the Isle of Wight, an interesting anomaly.

Figure 5.7   This cartogram of change shows hardly any change over these ten years. Change here in Britain would indicate such things as unskilled workers moving to live in 'better class' neighbourhoods and professionals moving to live nearer to where they worked. It is often just as important to discover that something is not happening as to find out what changes are occurring.

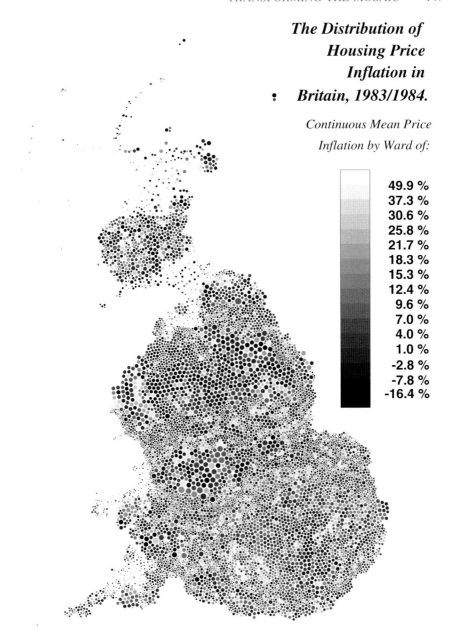

The Distribution of Housing Price Inflation in Britain, 1983/1984.

Continuous Mean Price

Inflation by Ward of:

49.9 %
37.3 %
30.6 %
25.8 %
21.7 %
18.3 %
15.3 %
12.4 %
9.6 %
7.0 %
4.0 %
1.0 %
-2.8 %
-7.8 %
-16.4 %

Figure 5.8   This is the first of a series of ward cartograms showing housing price inflation, all using the same grey scale. Housing prices, obtained then from building society records, can vary considerably by area, so this level of detail (10 444 wards) is needed to study it. In 1984 few sales were being made and recorded. You can see little more than largely random variation.

One outlying spot in ten thousand can easily occur. Far more notice should be taken when a clump is seen, for this is very unlikely to have arisen at random. Spatial and temporal smoothing could have been employed here, but it is remarkable how well the structure of local housing can be seen from the pictures of raw information. People are quite careful how much they pay for houses; it is a lot of money. Only maps of mortality rates are as smooth, as people are also extremely careful about trying to avoid a premature death.

Soon after 1983, a complex pattern begins to emerge of high inflation in the Home Counties and London, slowly moving out with a roughly ring-like ripple shape (Figures 5.9, 5.10 and 5.11). This picture mirrors the changes in unemployment described above. The increases become greater and greater, but a dark core begins to form in the centre, in the heart of London (Figure 5.12).

Suddenly the darkness appears and the house price slump of 1989 is upon us[11] (Figure 5.13). Only images like these could show how this began, preserving the detail, rather than averaging. In a few parts of central London prices fell in the year to 1987, more falls were seen in the very heart of London during 1988, then the whole South East turned black, but some had forseen the falls before they came. These market-savvy individuals were selling early and a little cheaply in the middle of the Capital just before the crisis. It is as if they had insider information.

After the late 1980s housing bubble burst the picture soon reverted to normal again. The static pattern for 1989 looked much like 1983, although the prices had trebled (Figure 5.14). These images should be borne in mind, however, in connection with other changes, and also in connection with how they relate to changes about to be shown. Unemployment and inflation are often claimed to be the major preoccupations of those who are about to vote: 'the economy stupid'.

5.6   Reshaping votes

The electoral change in February 1974 was therefore quite exceptional, not simply in magnitude but also in direction: the British pendulum stopped swinging.

(Crewe, Särlvik and Alt, 1977, p. 132)

In hindsight February and also October 1974 were key elections, points in time when decisions began to be made that reflected great changes in underlying trends, in this case changes towards accepting growing inequality, which began with a question over who ran Britain. The last time change had been as great

---

[11] 'Bramley and Paice ... have calculated that, even assuming that potential buyers can raise a 95 per cent mortgage on three times their income, one in three families living in the South East cannot afford to enter owner-occupation' (Hamnett, 1989, p. 111).

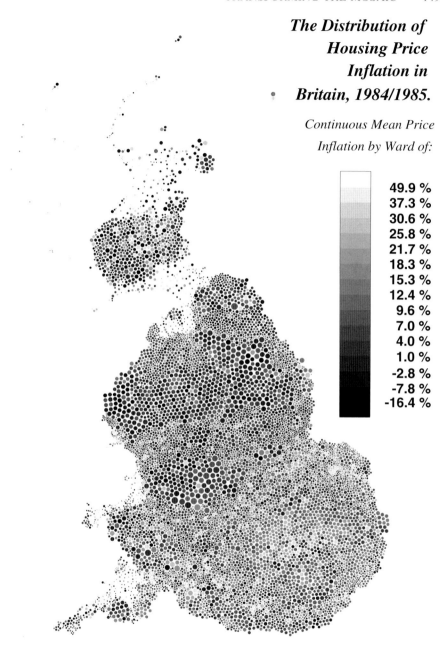

The Distribution of Housing Price Inflation in Britain, 1984/1985.

Continuous Mean Price
Inflation by Ward of:

	49.9 %
	37.3 %
	30.6 %
	25.8 %
	21.7 %
	18.3 %
	15.3 %
	12.4 %
	9.6 %
	7.0 %
	4.0 %
	1.0 %
	-2.8 %
	-7.8 %
	-16.4 %

Figure 5.9   Overall, this year the average inflation is around 7% but reached as high as 30% in a few places, particularly within central London. A clearer picture begins to emerge of the rising prices in the South, although there are still a few blackspots there where the prices are falling. Falls can be seen in the value of homes in central Liverpool and many places further north.

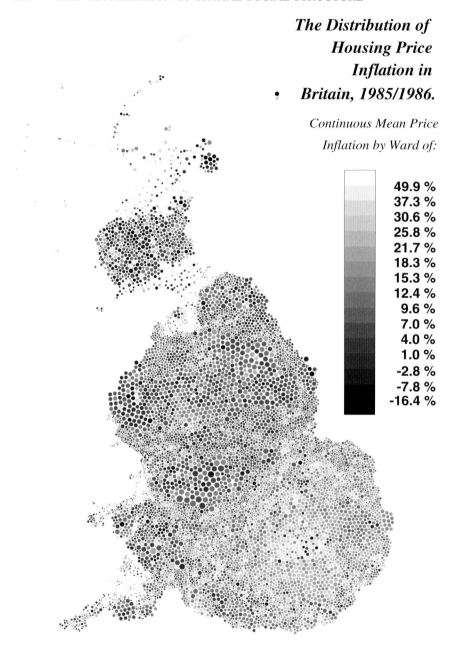

The Distribution of Housing Price Inflation in Britain, 1985/1986.

Continuous Mean Price Inflation by Ward of:

49.9 %
37.3 %
30.6 %
25.8 %
21.7 %
18.3 %
15.3 %
12.4 %
9.6 %
7.0 %
4.0 %
1.0 %
-2.8 %
-7.8 %
-16.4 %

Figure 5.10   Now most of the country shows an average of 9 % to 12 % housing price inflation. The picture becomes generally lighter, particularly in central London, although ironically some of the few darkest spots are also there. A fall has begun in Aberdeen. Its North Sea oil money runs drier, but in general a gradually increasing inflation rate is spreading out from London.

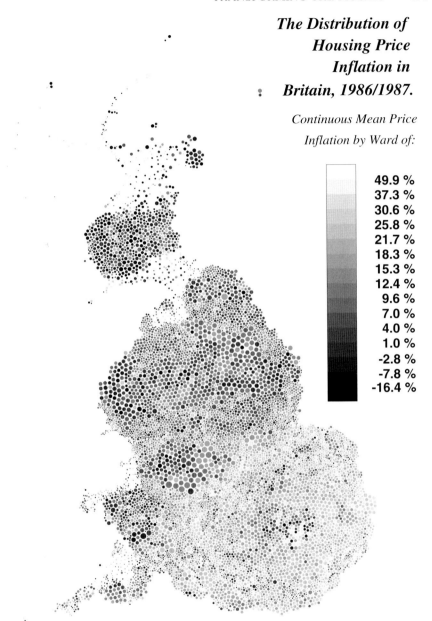

The Distribution of Housing Price Inflation in Britain, 1986/1987.

Continuous Mean Price
Inflation by Ward of:

49.9 %
37.3 %
30.6 %
25.8 %
21.7 %
18.3 %
15.3 %
12.4 %
9.6 %
7.0 %
4.0 %
1.0 %
-2.8 %
-7.8 %
-16.4 %

Figure 5.11   The average inflation has now risen to about 20%, often much higher southeast of a line from the Wash to the Severn estuary, where the picture has now become even lighter than before. In the London area annual inflation can be up to 50%, although even there in some wards prices fell. Elsewhere a dark ring has spread around Glasgow, while central Liverpool and Bristol are also not faring well.

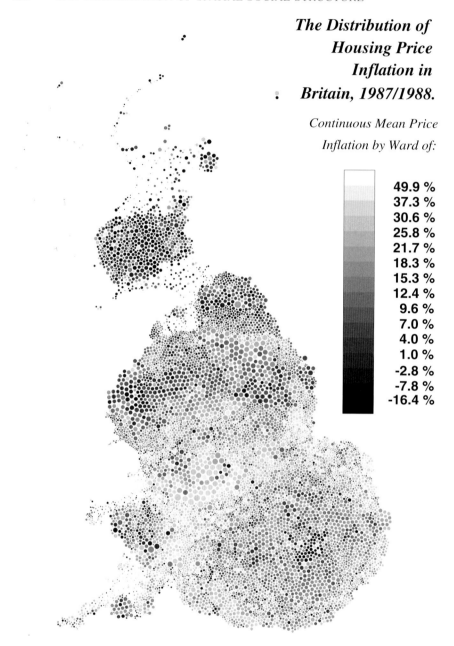

**The Distribution of
Housing Price
Inflation in
Britain, 1987/1988.**

*Continuous Mean Price
Inflation by Ward of:*

49.9 %
37.3 %
30.6 %
25.8 %
21.7 %
18.3 %
15.3 %
12.4 %
9.6 %
7.0 %
4.0 %
1.0 %
-2.8 %
-7.8 %
-16.4 %

*Figure 5.12   The housing bubble has spread out to cover most of Britain. This
is the lightest image of all. In five years housing prices have changed out of
recognition and many think they have made fortunes, but Liverpool and Glasgow
have been left out. At its centre, in London, housing price inflation appears to
have calmed down, but with some much darker spots in the heart of London.*

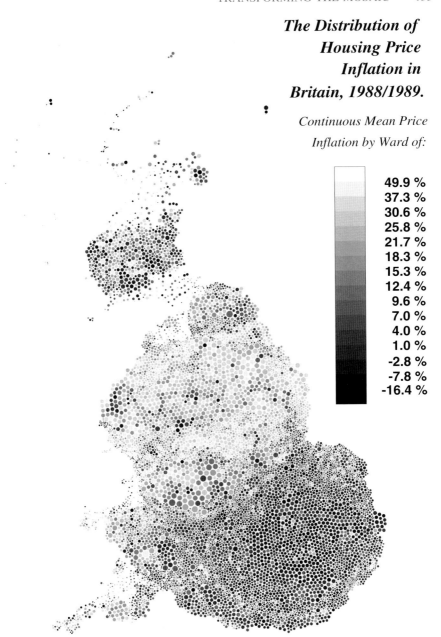

The Distribution of Housing Price Inflation in Britain, 1988/1989.

Continuous Mean Price Inflation by Ward of:

	49.9 %
	37.3 %
	30.6 %
	25.8 %
	21.7 %
	18.3 %
	15.3 %
	12.4 %
	9.6 %
	7.0 %
	4.0 %
	1.0 %
	-2.8 %
	-7.8 %
	-16.4 %

Figure 5.13   The picture has become dark over the entire southern portion of the country. The housing bubble has started to burst, but not yet everywhere. The picture in the north is now brighter than it has ever been and a reverse north/south divide has appeared. As housing prices start falling, many who bought recently will be plunged into negative equity, their homes worth less than their mortgages.

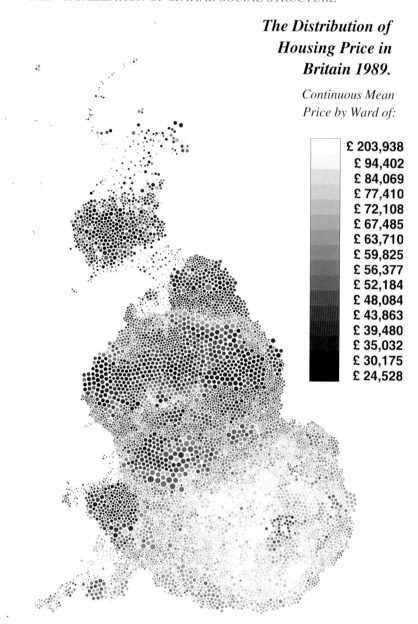

**The Distribution of
Housing Price in
Britain 1989.**

*Continuous Mean
Price by Ward of:*

£ 203,938
£ 94,402
£ 84,069
£ 77,410
£ 72,108
£ 67,485
£ 63,710
£ 59,825
£ 56,377
£ 52,184
£ 48,084
£ 43,863
£ 39,480
£ 35,032
£ 30,175
£ 24,528

*Figure 5.14   This picture looks similar to that for 1983 (see Figure 4.15). Some
lowest valued wards have shifted; in London, to nearer the centre. Prices in the
centre of Aberdeen have collapsed. However, look at the scales. At the bottom end,
average ward house prices have risen by 55%, at the top by 145% – fortunes
paid for by people going into negative equity and later by future house buyers
and renters.*

was in the early 1920s (especially 1923[12]). Those commentating were often most concerned with the issues of the day, but between these issues were other debates, questions of whether all people should be treated equally or if there were deserving rich and undeserving poor.

In considering voting in general elections, because they are held so infrequently it is necessary to go back in time before the era covered by most of this work to see how the changes of one period compared with other years. Ten general elections were held between 1995[13] and 1987 inclusive. By-elections have not been included here, but they have never altered which party became the government in power. What are of interest here are the changes that occur at all-out elections, those votes that change governments and how the political landscape slowly changed. Just as the housing market and economy goes from bust to boom and back again, those viewing these images want to be in a position to guess what they will see next. What was on view in the late 1970s and early 1980s was the building of the foundations for growing social and political polarisation.

Images can be produced showing which parliamentary constituencies changed hands between each contest (Figures 5.15 and 5.16). These are most important in showing the geography of political success. Only those seats that changed hands are coloured; on the outside of their symbols they are coloured by the colour of the former party holding the seat, inside by the new holders. Between three parties there are six possible colourings (between six different parties there are thirty). Here, the results of those changes, and something of what lies behind them, can be seen (Figures 5.17, 5.18 and 5.19).

Showing the changing proportion of votes is more problematic than simply showing who wins or loses outright. The British electoral system is dominated by three major parties. It is the swings between these which are of most interest.[14] The swing between three choices is a two-dimensional object (just as the simple swing between two choices, a basic change, is one-dimensional). A two-dimensional change can be shown in various ways. Some pictures (Figures 5.20 and 5.21) use arrows, the direction of which indicates the direction

---

[12] 'The February 1974 election was one of the most peculiar, and perhaps one of the most important, in British electoral history; ... It was called as a referendum on a specific policy issue for the first time since Stanley Baldwin did so – also unsuccessfully – over the tariff issue in 1923. As all know, the election not only stimulated the highest voter turnout since 1951, but also a mass exodus from both major parties – towards the Liberals in England and the Scottish Nationalists north of the Tweed' (Burnham, 1978, p. 280).

[13] 'It may be hard to believe nowadays, but during the 1955 general election campaign television news broadcasts made no references whatsoever to the election because the broadcasting authorities feared they would be in breach of laws regulating the conducts of elections' (Denver, 1989, p. 50).

[14] The ratchet was also seen to slip in both 1959 and 1983: 'The decline of loyalism within both major UK parties in the 1970s is well attested. Less obvious is the slip in support of 1959, linked to the resurgence of the Liberals in that year' (Budge and Farlie, 1983, p. 126).

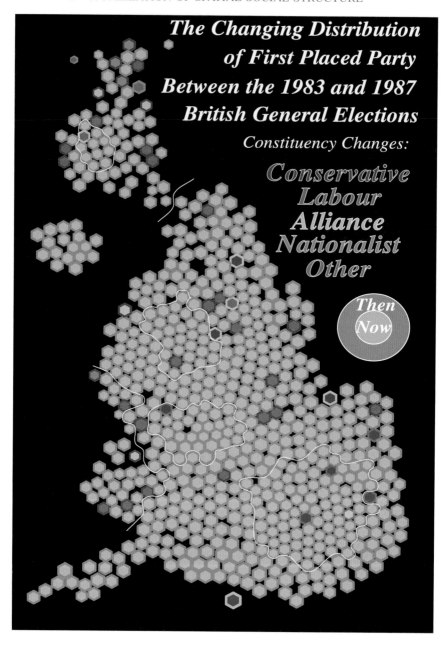

Figure 5.15   A simple glyph was used here to try and show which of 650 seats had changed hands – from whom to whom. The colour of the losing party is shown on the outside, the winning (more recent) party on the inside. The general distribution of changes is interesting, especially when not the general Conservative-to-Labour swing. There is a group of Labour losses in London.

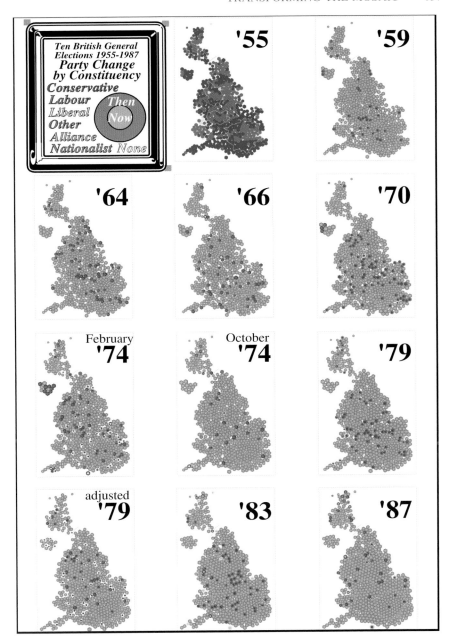

Figure 5.16   The shading in the previous figure is used here to show which seats changed hands between ten general elections from 1955 onwards. Following the 1979 redistribution of seats, white outer rings are used to show new constituencies. The changes are often un-patterned, but interesting clusters and other features can be seen, especially in all the Conservative Party gains in 1983.

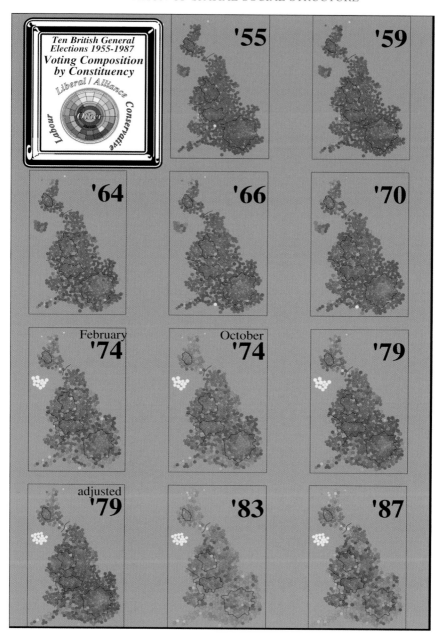

Figure 5.17   Here each constituency is shaded to show the proportion of votes given to each main political party. Northern Ireland goes its own way. The general change and increasing polarisation is evident. The ring formed around the Capital by 1987 mirrors a lot of the other distributions found in this book. Britain was politically a much less divided country in the 1950s.

of change while the arrow length indicates the magnitude of change.[15] Pointing up represents movement towards the Liberal party, left towards Labour and right towards the Conservatives. The advantage of the parties having conventional directions (left, right and centre) as well as colours (red, blue and yellow) has been exploited here. The arrows could be coloured by the existing shares of the vote, so as to show what political complexion the change was from.

What can be seen from these images of electoral fortune are uniform swings around the country with distinct variations depending on both geographical and political position. The divergence in political behaviour grew stronger, especially from 1974 to 1987, as the arrows in different parts of the country began to head in different directions, taking the voting compositions of the seats and the pattern of victories with them. The dramatically changing fortunes of the Liberal party are clearly shown.

The images shown here encompass two geographical redistributions of constituency boundaries, both increasing the number of seats being contested. These changes are incorporated in the graphics, the constituencies in which were sized by the electorates of their respective election, and so the cartograms change in shape on these pages. New seats squeeze in between their neighbours and old ones are squeezed out. All the time the general shape of the country is changing as people migrate both out of cities and, in general, to the South. The changing pattern of turnout is also telling (Figure 5.22). Like a living part of the social landscape, the political layers can quickly change their shape and colour.

5.7 Erosion and deposition

> ... London, at the heart of the South, has lost well over half a million manufacturing jobs during the last twenty years. To put this into perspective, London has lost almost as many manufacturing jobs as Scotland ever had. Indeed, some of the fastest-growing areas are found outside the traditionally prosperous South East and Midlands.
>
> *(Fothergill and Gudgin, 1982, p. 6)*

To understand why the City of London was deregulated in the 'Big Bang' of 1986, a move that is now blamed for the economic crash of 2008 being far greater than it otherwise might have been, requires knowing what came before 1986 and what that change was in reaction to. The British social landscape changes slowly but surely. The industrial infrastructure (the social geology) is most intransigent, but when it is altered all else must change. The population structure reflects movement and the changing fashions, like having children at particular ages and in particular numbers. The social classes into which children are born are the soil

---

[15] 'A feature of voting behaviour in Britain in recent years has been the increasing volatility of the electorate, with a growing proportion prepared to shift allegiance between elections' (Johnston and Pattie, 1989a, p. 241).

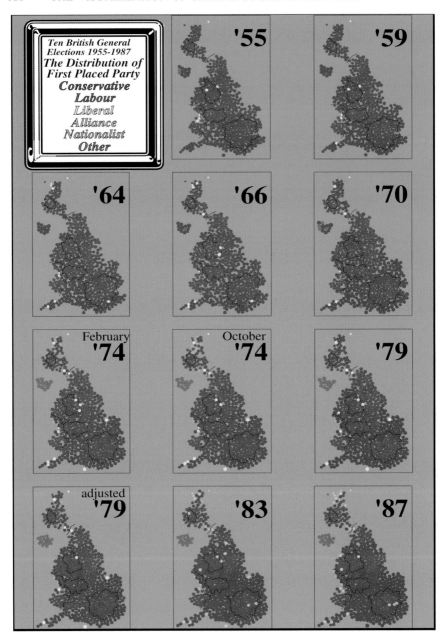

Figure 5.18   In comparison to the previous figure, these images simply show who won. Six distinct colours are used to represent the various victorious parties in 705 consistent constituencies over ten elections on the (changing) equal electorate cartograms. The Highlands have lost people while the South of the country has grown. Over time division became more intense – a sharper geography.

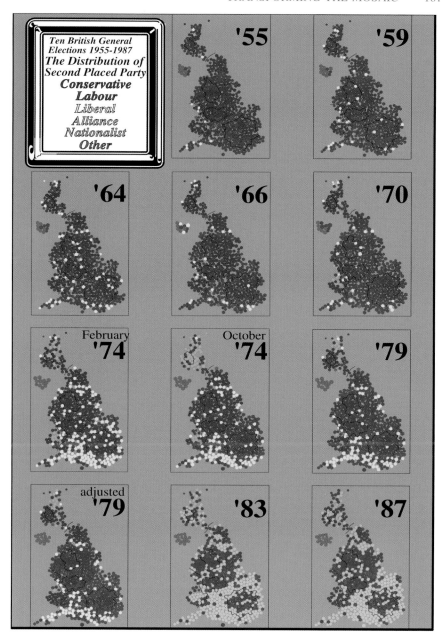

Figure 5.19   The change in the order of second place party over time is actually much more interesting than that of the winners. The geography of the rise of the Liberals (and then their Alliance with the Social Democratic Party) was dramatic, as they spread around the South. The rise of the Nationalists in October, 1974, was also outstanding and a glimpse of the future that was to come.

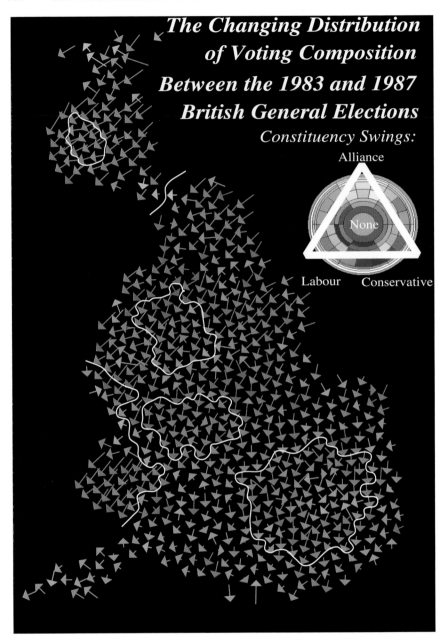

Figure 5.20   Arrows are especially useful when neighbouring groups point in similar directions. The colour shows the 1987 three party vote. The length of arrows shows the direction and size of the swings from the 1983 results. They can be seen to be flowing together, as the Labour strongholds move further to the left. The direction of movement is less certain in the Home County areas.

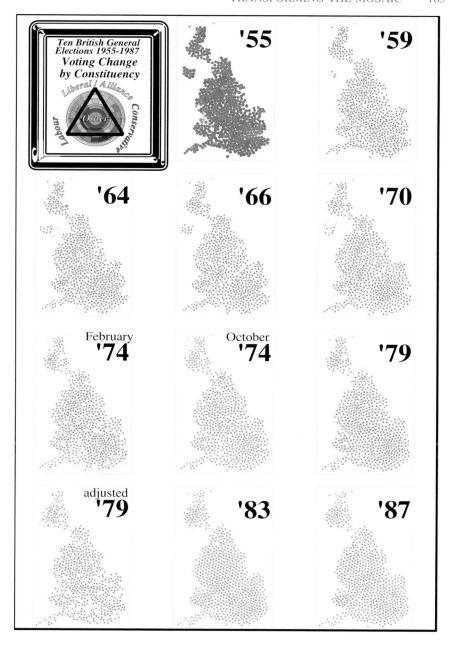

Figure 5.21   The first frame shows the distribution of votes in 1955. After that, this figure uses the same system of arrows as in the previous figure. Colour shows the mix of votes in each election, arrows show the direction and, by length, the magnitude of the change. Arrowheads are all the same size. The swing changes from uniform to become more polarised.

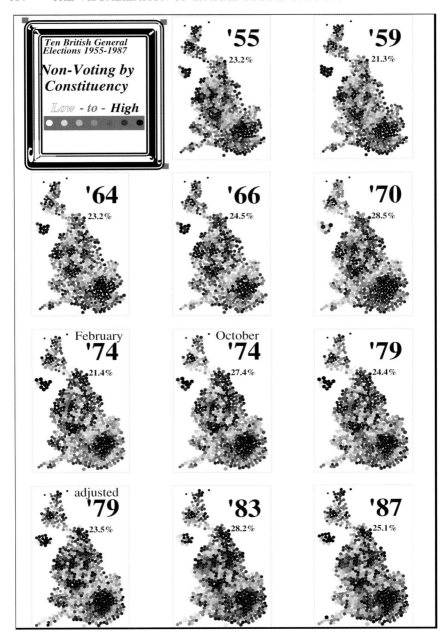

Figure 5.22   The pattern of abstentions by registered voters changes slowly. The North East of England has now a lower turnout and that in Northern Ireland rose and then fell. The area of the West Midlands that contains many people who do not vote has also swelled in size. Overall the average turnout has been between 71 % and 79 %, tending to be higher locally in marginal seats, but sometimes also in safe ones.

of our social landscape and come in many colours. The colours become more distinct as society polarises.[16]

The changing positions of lifetime migrants show several aspects of economic restructuring through the movement of people, from where they choose to live and to where they are perhaps constrained to reside. The living parts of our landscape, those that change day-to-day, change fastest. The distribution of jobs, of wealth, of housing and of how people vote depends upon all the other layers in the overall picture and on each other to an extraordinary extent.[17]

How the landscape is changed depends on many forces. It is the flows of people, like flows of water, that both maintain and alter the picture. Every day, the flow of people to work links industries to population; then people suddenly move in a quite different way: they move to different homes – they migrate. It is the streams of migration that sculpture our landscape, transforming its structure, depositing a new workforce and eroding the old. These flows of people, which maintain and change the social landscape, are the subject of the next chapter of this book, which is about ways of seeing into the continuous spatial movement that makes social structure. Humans are always moving.

---

[16] '... these trends suggest it is not impossible to envisage the development in the not too distant future of a socially polarized inner London, divided by tenure, with middle class owner-occupation juxtaposed with a residualized and predominantly working class public rented sector. Those groups excluded from this process may be displaced into outer London where many inner suburban areas may become transformed into lower value ownership mixed with much of what will be left of the private rented sector. The net result therefore will be a stabilising but polarized inner city and a declining suburban ring. In the process "inner city" problems may become gradually displaced into the suburbs' (Hamnett and Randolph, 1983, p. 164).

[17] As the former Capital of a long-gone empire London had suffered greatly. A year after the banks were deregulated to slowly start growing again (but now more on the profits of gambling) it was being said still that: 'In 1987 London's economy is in deeper crisis than it has been in for a hundred years. In certain clear respects it is worse than the 1930s. The rigours of those depression years affected other regions more than London and did not bring quite the same extent of misery and insecurity to the Capital' (Townsend with Corrigan and Kowarzik, 1987, p. 12).

6

Cobweb of flows

Look upon the population and its various activities as a part of a vertically-rising stream in space-time with oblique tributaries of movements in a short chronological perspective and a longer one (for example, daily journeys to work and migratory moves respectively).

(Szegö, 1987, p. 200)

6.1 What flow is

Flow is more than change. It describes the static structure of change: how the change came to be, what changed. Change is a difference; flow is an entity in itself. It is of an order of complexity above change and involves a quantity of information of an entire order of magnitude greater than change (something reflected in the file sizes of the images shown). In our social structures it is the movement between places, rather than alterations within them, that are responsible for the restructuring we see.

To paint a picture of the static social structure at any point in time, we need only know the situation in each cell of the structure. To show where that structure is changing we need to know the situation around at least two points in time for every cell. To show how that structure changed, what moved from where to where, information is required about the relationships between all cells (Box 6.1). Flows are real. The people they measure did move from one place to another (Figure 6.1). This is the structure of change, the structure of movement. It forms a cobweb that links places.[1]

[1] Most census mapping is of the distribution of night-time population: 'This effect, which sometimes distorts at least some aspects of city life, can be overcome in some degree by measuring changes over time and movements in space; the static structure is only a departing point for the analysis of a living city' (Shepherd, Westaway and Lee, 1974, p. 112).

Box 6.1   Storing the flows

The full matrix of travel-to-work flows between the 1981 British wards contained in theory $10\,444^2$ or over 100 million separate counts. It was not possible to obtain flows within Scotland from the census records available. Only 434 340 or half of one per cent of all possible routes were actually travelled, by 20 602 790 people, as people work reasonably near to where they live. The entire matrix of commuting flows was stored as a run length encoded binary file of only 628 752 bytes in length. This was achieved by sorting the flows from each ward in ward order and recording only the displacement in ward number and size of the flow, each in four bits. A carry option could be set if this size of field was inadequate. Thus the set of flows from any ward could be determined instantly and the entire matrix easily held in computer memory.

The migration streams were more dispersed and also subdivided by sex. 4 210 900 people moved across ward boundaries between 1980 and 1981, resulting in 893 941 migration streams. It was stored in a file of 1 453 252 bytes (which would just fit on a floppy disk).

The final step in complexity taken here is to look at the change in flows. Again, the information required at least doubles: the flows between all pairs of cells at several points in time. The measure of how many more or less people moved between two places from one pair of years to another provides an abstract quantity, not easy to comprehend[2] (Figure 6.2).

6.2   What flows there are

> *In a time–space region, each individual can be visualized as a contin-uous path starting in a point of birth and ending in a point of death. Depending on the observation period, individual-paths can be referred to as day-paths, year-paths or life-paths. This corresponds to the concept of life-time in demography ....*
>
> *(Carlstein, 1982, p. 41)*

Many flows are not spatial. The flow of votes between political parties produces a small matrix within each constituency or ward. Each element of the matrix tells us how many people changed their vote from one party to another, or chose not to vote. Often flows are not given; only net change can be derived.

---

[2] It is the streams of migration that are real and of importance – not the marginal distributions they produce: 'The simple framework that reduces migration to "pulls" towards desirable locations and "pushes" away from less desirable ones cannot adequately explain even total migration, let alone the spatial structure of streams' (Mueser, 1989, p. 196).

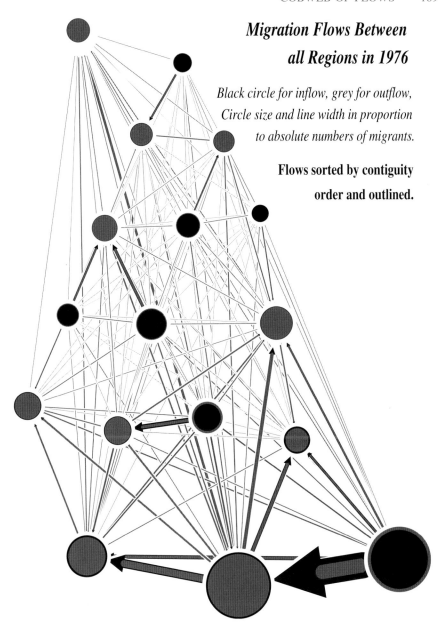

Migration Flows Between all Regions in 1976

Black circle for inflow, grey for outflow, Circle size and line width in proportion to absolute numbers of migrants.

Flows sorted by contiguity order and outlined.

Figure 6.1   Flows of people are sorted so that arrows between neighbouring places appear uppermost. This is a particularly useful technique that allows movements between specific places to be more clearly seen. However, by including just seventeen places, the picture shows much less than before (see Figure P.4 in the Preface). Would the picture have been much clearer if just a particular age group could be shown?

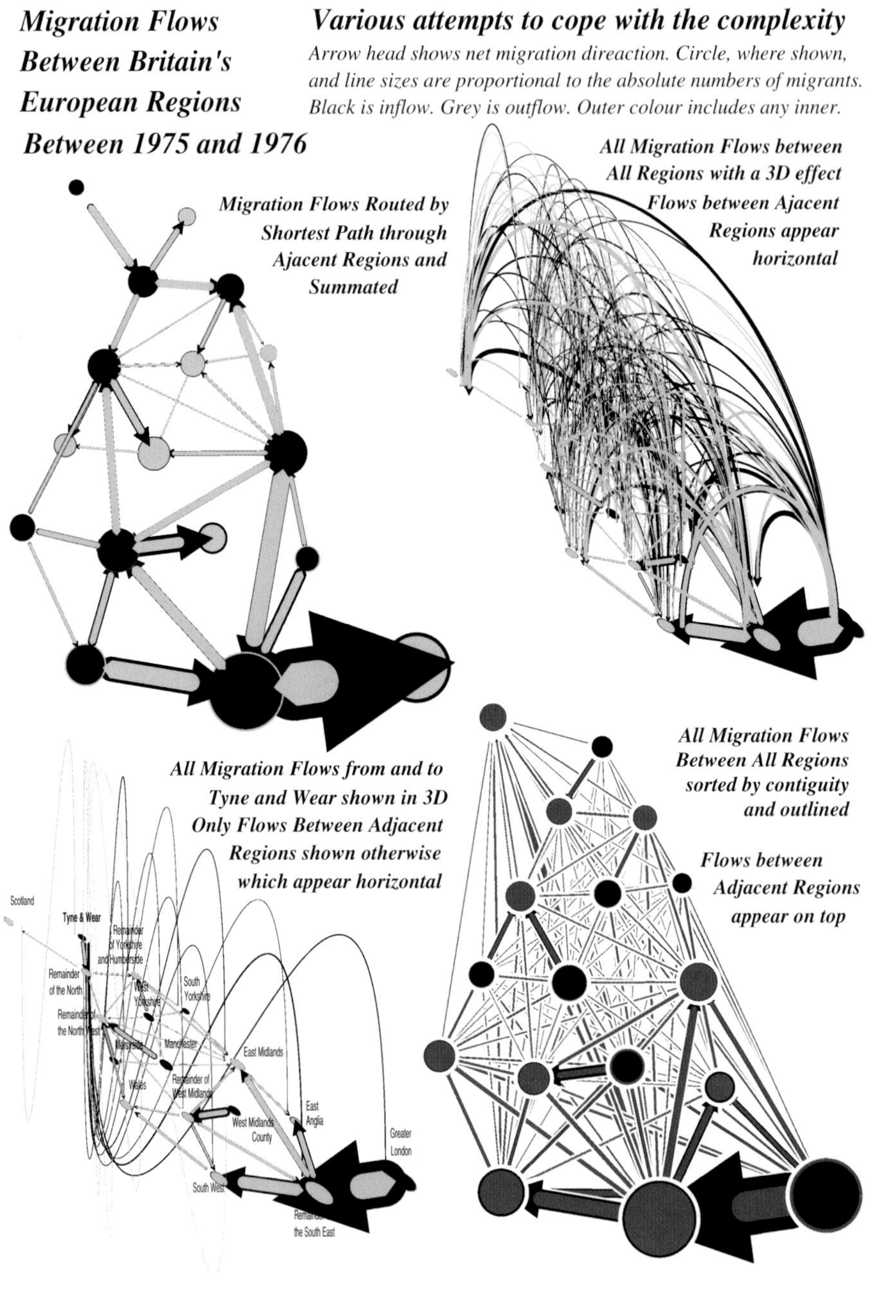
Migration Flows Between Britain's European Regions Between 1975 and 1976
Various attempts to cope with the complexity
Arrow head shows net migration direaction. Circle, where shown, and line sizes are proportional to the absolute numbers of migrants. Black is inflow. Grey is outflow. Outer colour includes any inner.
Migration Flows Routed by Shortest Path through Ajacent Regions and Summated
All Migration Flows between All Regions with a 3D effect
Flows between Ajacent Regions appear horizontal
All Migration Flows from and to Tyne and Wear shown in 3D
Only Flows Between Adjacent Regions shown otherwise which appear horizontal
Scotland
Tyne & Wear
South Yorkshire
Manchester
East Midlands
Wales
West Midlands County
East Anglia
Greater London
South West
All Migration Flows Between All Regions sorted by contiguity and outlined
Flows between Adjacent Regions appear on top

Figure 6.2 *When flows are artificially routed through areas it looks as if far more people move to London (top left diagram). Don't route them, but drop them in as curved lines (top right) and it is clear more people leave London for the South East. Simplify that map and you lose too much detail (bottom left). None of these are an improvement on the previous figure, also shown here (bottom right).*

Flows on and off the unemployment register are an aspatial set, which are recorded. Here, within every employment area, only two states are given, moving on or moving off the register. With a long time series, however, these little matrices can tell us a lot about the frequency and average length of unemployment.

Flows of spatial movement abound and are usually categorised by purpose of movement. Some have been mapped in the past, but we have little information about many others: frequent flows to go shopping or to school for instance. Less frequent, but important moves, to holidays, hospitals or further education, for example, are also difficult to obtain data for or to estimate.

The only spatial flows regularly recorded and disseminated are those between home and workplace, and migration from one home to another. Travel-to-work flows are enumerated at the census (Figures 6.3 and 6.4). They link the basic static points at which people are regularly enumerated; they give day and night time populations and tell us how these change.

Migration is counted at a fine spatial scale by the census. A useful series can be obtained from the National Health Service central register of patients. This links everyone's NHS number to a specific General Practitioner and so records when they move home as they tend to move doctor too. Migration flows have been collected for many years to see how the night time population is changing in the medium term. The creation of flow matrices was a by-product of this.[3] Here all this is depicted as if a cobweb of migration was wrapped around a honeycomb of social structure. The honeycomb is the cartogram. The cobweb is the flows.

6.3   Unravelling the tangles

> *If the movement from each state to every other state were indicated separately, the multiplicity of lines would have made the map totally illegible, but through combinations of migrations in the same direction it has been possible to preserve legibility and still to show what was intended.*
>
> *(Thornthwaite and Slentz, 1934, p. 14)*

Places in Britain can be divided into two: those where more people leave for work in the morning than arrive to work (these are residential areas) and those where more people arrive for work than leave (industrial (work) areas). Such a map could have two colours, black and white. It would show us the rings and sections of the basic commuting pattern. But how much of what we could know would this very basic picture tell us?

---

[3] Deviation from the expected propensity to migrate is mapped in some of the illustrations in this book: 'To discover any flows of unusual magnitude between certain areas, we calculated expected migration flows. These prognoses are derived from the total out-flow from one area and the total in-flow to another. (These form the expected cell frequencies that are also determined in chi-square analysis.) The expected frequencies are subsequently compared with the observed frequencies (actual flows). The difference reveals any unusual attraction or repulsion exerted by an area' (Jobse and Musterd, 1989, pp. 247–248).

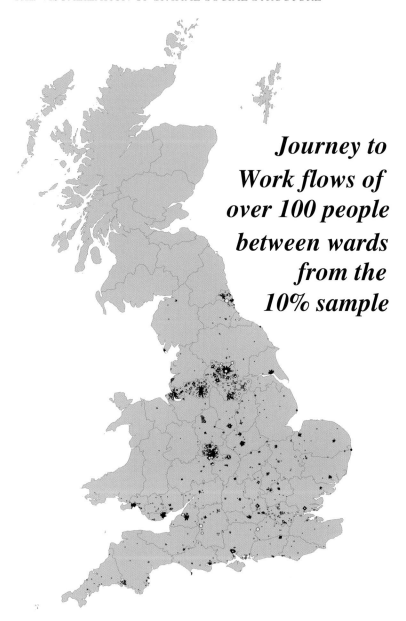

Journey to Work flows of over 100 people between wards from the 10% sample

Figure 6.3 Unlike house moves, journeys to work all have the same purpose. Here, only large flows between the 9289 English and Welsh 1981 census wards are shown on a land map. Arrows are visible for the larger flows. Most major towns are easily seen. Figures are from a 10% sample of the population, excluding Scotland. The yellow dots and circles show where many people live and work in the same ward.

Journey to Work flows of over 70 people between wards from the 10% sample

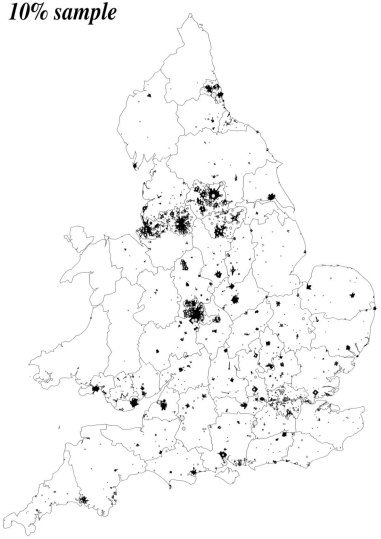

Figure 6.4   Large and medium sized flows between the 9289 English and Welsh 1981 census wards are shown here on an equal land area map. The arrowheads and widths are in proportion to the flow size. Surprisingly, London still does not show up because there are too few flows between unique pairs of wards out of and into London at this scale. The London commuters are all going home in too many separate ways.

Elaboration can begin by moving from two colours to a continuous colour scale. The image is smoother. However, just how many people are moving? The proportion of people moving out of each place can be shown, or that of those moving in, or both. This can be done with the colour of the cells, or their size, using size for indicating the proportion involved and darkness for the proportion outgoing, for instance.

Next, consider where, in general, people are going to and coming from. This is implied by the urban, suburban and rural rings of similarly shaded areas in the map being described here, but the patterns may not be as simple as we imagine.[4] Arrows can be drawn showing the average direction of outgoing movement by changing the shape of each cell to be an arrow, one that points to where the people are travelling (Figures 6.5 and 6.6).

After direction we can consider how far people are moving or, better still, how long it is taking them to get there. The length of the arrow is another aspect we can alter. The length is in proportion to the average time it takes people to travel; the direction indicates where they go, the size shows how many of them go and the colour – say through its level of saturation – depicts what proportion is outgoing.

The succession of images just described has been concerned with average flow. Such maps are useful, for instance in studying commuting, because most areas are either residential or industrial and most people go in *roughly* the same direction and travel the same amount of time. It is only the nature of travel to work that allows it to be mapped like this (Figures 6.7 and 6.8). The arrows could not be used to show where people were coming from or how far they travel to get to the shops, for instance, as these aspects are too variable, being concentrated in the centre of cities and of large neighbourhoods. However, it is better to show them as straight lines rather than flows along winding roads.[5]

The images shown here highlight the most important flows, but many flows are not shown at all.[6] The patterns seen are partly an artefact of the precise

---

[4] '... in some cases, the number of in- and outcommuters are almost the same so that they should be represented by the same size of circle. If the same size of circle is used for each, there is no rim left to indicate which group predominates. If it is attempted to show a rim, then the graded circles are not accurate. The main advantage of this method, however, is that in- and outcommuters can be represented on a single map' (Dale, 1971, pp. 17–21).

[5] 'Tracing the actual itineraries is not sufficient for representing a system of relations. A map of maritime routes, even when weighted, does not show the direction of trade among the centres of activity; it shows the density of ships at sea. The maritime trade among the cities of Europe and the Mediterranean will only appear in its diversity, weight, and geographic direction, when each connection, even through maritime, is represented by a straight line ...' (Bertin, 1983, p. 344).

[6] 'The cartographic difficulties involved prevent the representation on one map of all net movements from each state to every other state, yet, since it seemed necessary for purposes of comparison to get as much of the movement as possible on one map, it appeared desirable to affect a certain compromise. It was found that ... the optimal threshold for the deletion of entries is the average flow size. This data selection rule deletes as much as 80 % of the flow arrows, but generally only 20 % of the flow volumes' (Tobler, 1987, p. 155).

Journey to Work flows of over 20 people between wards from the 10% sample

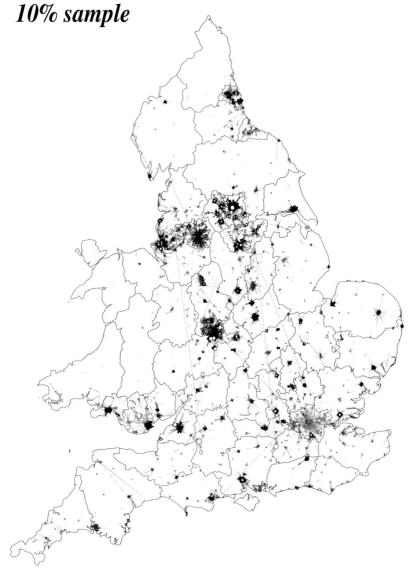

Figure 6.5   Small as well as larger flows are shown on this land map, giving the actual distance to work. Circles are where many people live and work in the same ward. Flows of over 70 people between pairs of wards are shown by thicker black arrows whose heads and widths are in proportion to the size of the flow. Smaller flows are shown by thin grey lines. Flows into and out of London are now visible.

Journey to Work flows of over 10 people between wards from the 10% sample

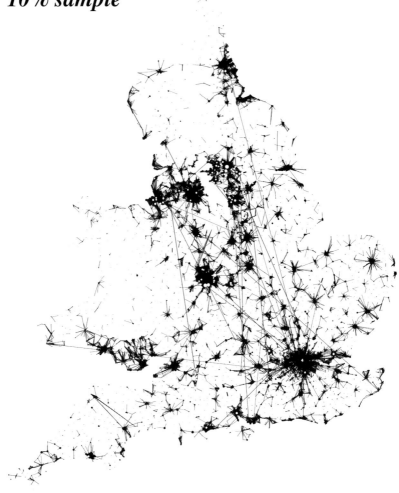

Figure 6.6   Most flows are shown between the wards on this land map. The patterns of people moving can be studied in great detail, but the magnitude of flows cannot be compared. All the lines have to be thin to be seen. Here less obvious features can be made out, such as the interconnected centres east of the Pennines from Leeds to Nottingham. Some extraordinarily long commutes become visible.

method chosen. From seeing these pictures you could think you might draw lines dividing the headwaters of the flow to define travel to work catchments. Whether that is possible or not depends on how much flows cross over, how many people travel in the same direction, how you measure flow and on how many people travel at all.

6.4 Drawing the vortices

All the variations introduced by spatial, numerical and temporal aggregation procedures operate on origin and destination data in an almost more bewildering variety than they do on static data. . . . we summarise by trying to pick out the main bundles of arrows moving between pairs of areal units. Thus it depends entirely what size, shape and position of spatial units we use, what apparent bundles we pick up.

(Forbes, 1984, p. 99)

The objects that have been discussed up to now have been vectors, single attributes for sets of single areas – average direction and magnitude of flow. These are all simply summaries. It can be better to attempt to draw the matrix, not just a tangle of lines connecting all places between which people flow but a picture that shows the static structure of the change in as much detail as possible.

Lines can be used to show matrices of flow, as areas can be used to depict scales of attributes. Lines have length, width, colour, direction (indicated by arrowhead), and can be given order as they overlap. The first of these qualities, length, conveys the strongest information, but is most difficult to use, as the line links two places, and to alter its length would make that connection ambiguous.

The best choice we have, which has most influence over the final image, is whether to draw any particular line at all (Figures 6.9 and 6.10). With areas, this is not a choice; there is a place for every area. With lines, there is only enough space for a minority to be shown (between more than a couple of dozen places). A line should be drawn between two places only if the flow between those places is significant. But what is a significant flow?

There is a second and linked difficulty to determining which flows are worth showing; this is the most serious effect of what some geographers call the modifiable areal unit problem: the problem that the boundaries of each area influence any statistics of that area. For mapping a simple variable the problem is largely overcome by the use of many small areas and by using cartograms of the denominator of the rate to be mapped. The problem is that the shapes, on the ground, of the arbitrary areas between which flows are recorded will affect the number

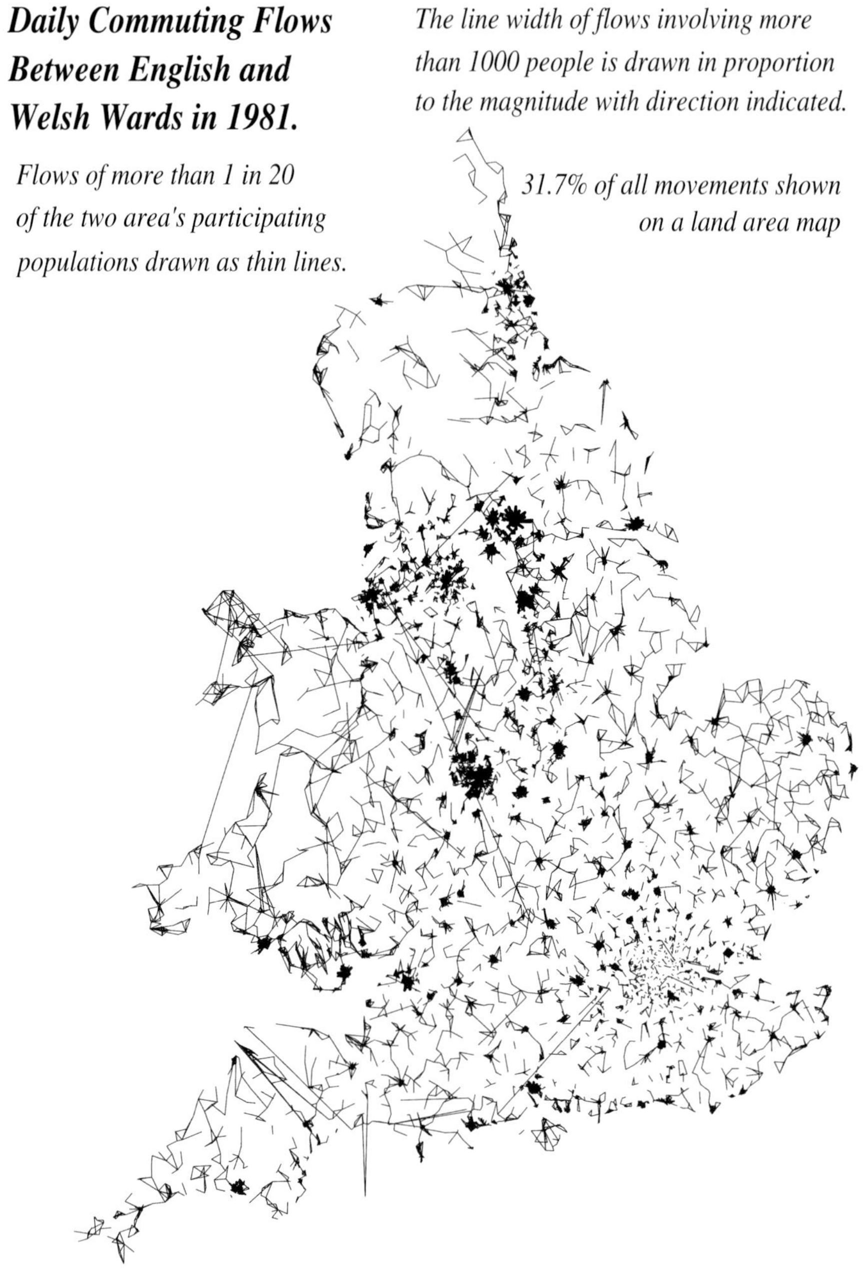

Daily Commuting Flows Between English and Welsh Wards in 1981.

Flows of more than 1 in 20 of the two area's participating populations drawn as thin lines.

The line width of flows involving more than 1000 people is drawn in proportion to the magnitude with direction indicated.

31.7% of all movements shown on a land area map

Figure 6.7   As the criteria are relaxed for deciding whether a particular flow should be included, more and more lines can be drawn on the map and soon it begins to appear as if every other town is connected to another town through one odd flow stream or another. Drawing flows of much greater numbers of people as thick arrows, not thin lines, adds a little structure to the map.

Daily Commuting Flows Between English and Welsh Wards in 1981.

The line width of flows involving more than 1000 people is drawn in proportion to the magnitude with direction indicated.

Flows of more than 1 in 20 of the two area's participating populations drawn as thin lines.

31.7% of all movements shown on a population cartogram.

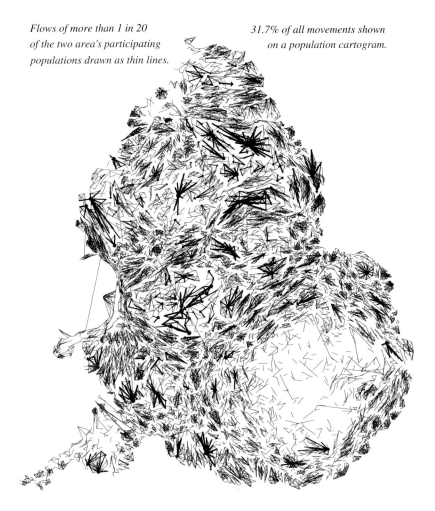

Figure 6.8   On the cartogram the flows shown on the previous equal land area map now form distinctive clusters, showing how each settlement is a separate entity in space. The thicker lines give greater definition to particularly well-focused urban centres. Lengths of lines here reflect commuting time more than distance. Most flows out of and into London wards are too diffuse to show up.

*__Daily Commuting Flows
Between English and
Welsh Wards in 1981.__*

*The line width of flows involving more
than 1000 people is drawn in proportion
to the magnitude with direction indicated.*

*Flows of more than 1 in 30
of the two area's participating
populations drawn as thin lines.*

*43.6 % of all movements shown
on a population cartogram.*

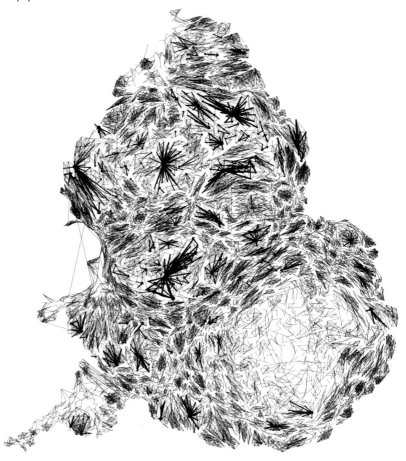

*Figure 6.9   Reduce the threshold for whether a flow should be shown and, on the
cartogram, the towns and cities become even more clearly defined. Boundaries
could be drawn around these clusters of lines to delimit travel-to-work areas in
which, in various ways, a majority of people both lived and worked. Borders particularly distinct at this time are clear, for instance between Leeds and Bradford.*

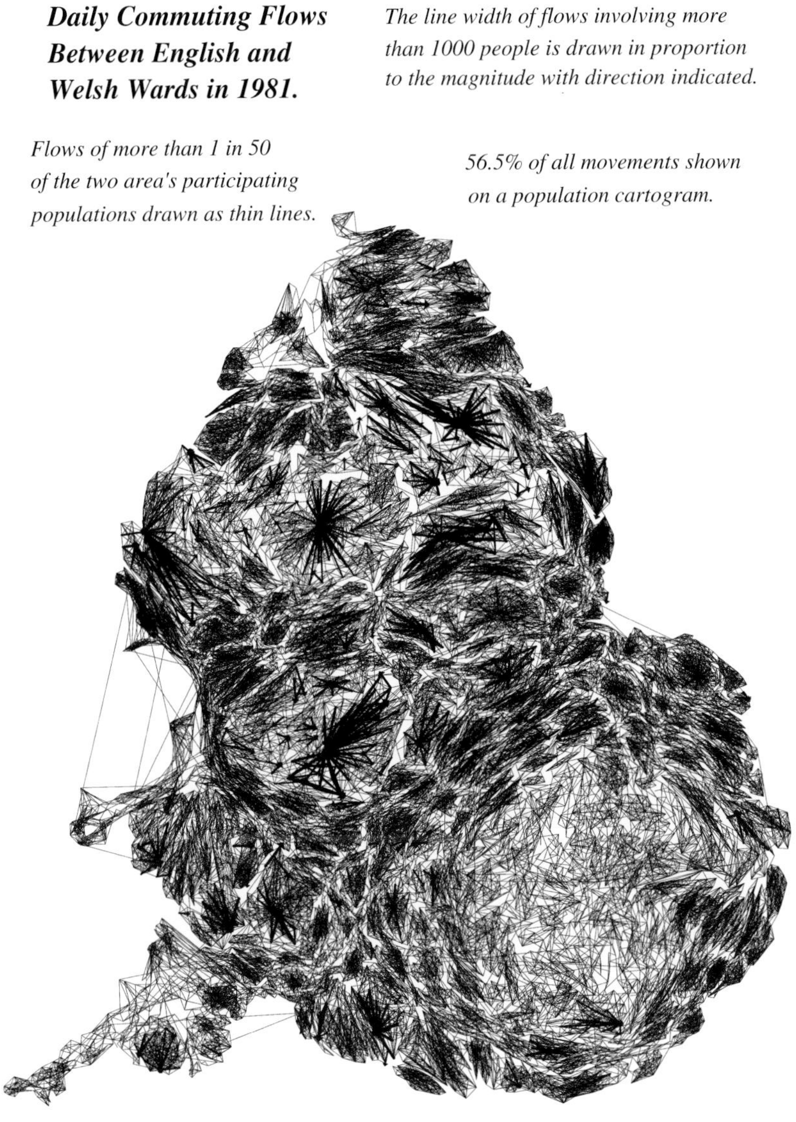

**Daily Commuting Flows
Between English and
Welsh Wards in 1981.**

*The line width of flows involving more
than 1000 people is drawn in proportion
to the magnitude with direction indicated.*

*Flows of more than 1 in 50
of the two area's participating
populations drawn as thin lines.*

*56.5% of all movements shown
on a population cartogram.*

*Figure 6.10   Reduce the threshold further for whether flows should be shown and
the towns and cities of the North-East of England begin to coalesce. Manchester,
Leeds, Bradford and Birmingham remain distinct. Within London, and particularly on the edge of London, clusters can begin to be seen forming, highlighting
structure similar to, but far more diffuse than, that found outside that city.*

Box 6.2 A significant flow

A flow between two places is of a significant magnitude if it is larger than you would expect, given the populations of the two places and the general propensity to move between places. Here a flow was deemed to be significant, and thus drawn when

$$\sqrt{\frac{m_{ijst} m_{jist}}{P_{it} P_{jt}}} > \frac{\sum\limits_{i}^{n} \sum\limits_{j}^{n} m_{ijst}}{(n-1) \sum\limits_{i}^{n} P_{it}}$$

where m_{ijst} is the flow from place i to j between times s and t, and P_{it} is the population who could move at place I at time t (n is the total number of places being considered). The equation therefore calculated whether the geometric mean propensity to move between two places is greater than the expected propensity to move.

Problems can occur when this method is applied to commuting flows, particularly in central London, where many people are moving into an area of generally low night-time population. It can be useful to use the day-time population estimate (caused by the flow) as the denominator.

of flows counted. This can have as great an effect as does the number of people who are actually moving.[7]

More people will flow from larger areas, but less people will move between such areas (since more move within them; see Box 6.2) and more will flow from long narrowly shaped areas than from more compact places. Put simply, the longer and wigglier the boundaries are over which flows are recorded, the more will be recorded.[8]

One solution to the problems raised here is the same as earlier: to use as many small areas as possible and to record as many of the flows that occurred as possible. Here wards represent the finest resolution. With flows, however, long distance movements can come to dominate the image, stretched across many other areas and perhaps visually concealing more important flows. Two solutions are used – one visual, the other statistical.

---

[7] 'Migration is an event which by definition involves two places – even if only adjoining houses. So spatial location and spatial units are more basic to migration than to other events such as births, deaths and consumers' expenditure. For these latter events, space is not fundamental to the event itself: it is merely necessary to define the geographical limits of the population to be included...' (Craig, 1981).

[8] '... for areas of the same size out-migration rates would be higher for a long and narrow areal unit than for a circular one' (Duncan, Cuzzort and Duncan, 1961, p. 34).

Box 6.3   Drawing overlapping arrows

This diagram shows how the sizes of migration streams between two places can be shown by a single arrow.

The arrowhead points in the direction of net flow, the size of the larger flow being shown by the width of the black line and the minor flow by the width of the grey line placed within it. A white border is placed around the arrow to separate lines lying beneath it and to clarify the image. The order of the arrows is then as follows: flows between neighbours are uppermost fol-

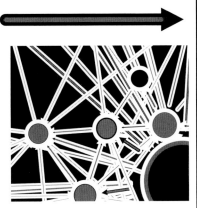

lowed by second, then third order contiguities. The inset shows part of the country level migration structure north-west of London.

Firstly, the flows can be ordered; lines can be placed above and partly obscuring other lines. This is achieved by drawing a slight white border around the lines. The strongest flows could be uppermost, but it was found most advantageous to put those that were between contiguous areas foremost, those of second order next and so on. The effect of this is to clarify the image, hide the most obscure deviance and show the most structure. Only strong unusual movements will show through the mesh (Box 6.3).

Secondly, flows can only be drawn that represent more than a particular type of average propensity to move.[9] The proportion of an area's population travelling to another area to work must be greater than the national average propensity for the line to be drawn. This usually means that over three-quarters of those travelling are represented on the image and that these are the most typical commuters.[10] This technique is particularly useful when flows between

---

[9] 'For all but the simplest networks these link data displays have many intersecting lines and are difficult to interpret. There are several possibilities for reducing clutter. One is to shorten the line segments, that is, instead of drawing the line segments 50 % of the way between nodes, draw them 30 % or 10 %, say. Another is to draw only lines whose corresponding statistic falls above or below some threshold. The difficulty with these ideas is that it is quite hard to come up with a good heuristic for setting these thresholds or line lengths (or overall line thickness, for that matter) before making the display' (Becker et al., 1990b, p. 93).

[10] It is worth remembering that it is these flows that keep the static structure stable, not just commuter flows maintaining cities, but migratory flows maintaining that nature of neighbourhoods. Most flows reinforce, rather than change, the structure: 'Many movements of individuals in the population do not alter the characteristics of the area, since one council tenant often replaces another, and one stockbroker moves only to sell his house to another stockbroker and so on. Even where an atypical individual arrives in an area, his new environment may influence him towards adopting

all the wards of Britain are being shown, as there is too little space to use the visual ordering technique effectively or to draw anything other than all thin lines (Figure 6.11).

Width can be used to convey magnitude of flow and arrowheads to convey the direction. The fewer the number of areas between which flows are being drawn, the more effective these methods can be. Colour can be reserved to show other things about those who are travelling: what kind of areas they are coming from or going to and what kind of people they are. The length of the line is a good surrogate for the time taken when the population cartogram is used as the base. It tends to take longer to pass by more people. This base also makes the picture clearer and gives some meaning to the density of lines (Figure 6.12).

So far, by looking at a relatively constrained set of flows we have not encountered all the problems this kind of mapping can create. It should be noted that the picture produced is very sensitive to the denominator used (Figures 6.13 and 6.14). Next, by adding colour, more can be shown. Then, by considering also migration, another web of long term movement can be mapped upon the first diurnal flows.

6.5   Commuting chaos

Programming the mapping of other than the most crude versions of such flow maps is not trivial, not least because large flows tend to occur between areas close together on the ground and numerous lines occur if all flows are mapped. Most frequently, they are mapped by arrows whose width is proportional to the flow involved.

(Rhind, 1983, p. 176)

What the commuting flows show us is the well-known city structure of Britain and the extent to which this is a true interpretation. On the population cartogram the flows appear much more confused than it is suggested they are when plotted in supposedly normal (Euclidean) space. This brings the image closer to the messiness of reality,[11] but in reality only a minority of travel flows are of commuters.

---

the characteristics of his new neighbours. Correlations and slopes between 1966 census data and partisanship change so little when using election years other than 1966 that we would be surprised to discover large biases due to the ageing of the census' (Miller, Raab and Britto, 1974, p. 399).

[11] Visually, ordering lines according to spatial proximity relieves some of the confusion: 'There is a particular problem in using line segments to connect locations on a map: the geographically longer lines are visually dominant. In many data sets, such as migrations or trade flows, the flow between locations varies inversely with distance. One consequence is that, at times, short but important lines may be difficult to see because of the long, less important lines. One possible way to alleviate this problem is to use wide lines for connecting nearby locations' (Becker *et al.*, 1990a, p. 289).

Many adults are not paid to work; in many areas these are the majority. Many of those who do paid work do so within the ward in which they live. Thus even at ward level commuter mapping is not representing the majority of people. Circles, their areas in proportion to the number of people living and working in the same ward, can give an indication of this phenomenon, while the flow lines between wards are still shown.

At this resolution the direction of flow is implied from the context – where the lines converge. Where lines cross it can be seen that the directions differ. There is a problem when two major commuting flows overlap in completely the opposite direction, but this is rare. The magnitude of flow is also difficult to see as the lines all appear very thin. They are thin partly because almost all flows are small.

Colour can be used effectively to show how the structure of flow relates to other social structures (Figures 6.15 and 6.16). The lines can be coloured according to a particular feature of the areas: where people leave from to then travel to work, about the work they do, where they are going, or even all three aspects simultaneously.

Finally, it would be possible, though perhaps too confusing, to place the flow map over a smoothed three-colour cartogram of some related aspects of the population involved. If this were done with migration, the cartogram could represent a few of the changes that the flows were producing.

The flows of travel to work are the heartbeats of the urban system, people being pumped in and out of the cities. If they stopped there would be no cities. The rhythm is well known, five days a week until Christmas. It may miss a beat, but not often. Seeing the flows is fundamental to understanding them. Understanding them shows us how society ticks.

6.6   Migration networks

The fact that among men aged under 50 the percentage of migrants who were unemployed was greater for those moving within local authority districts than for all migrants is difficult to reconcile with the hypothesis of movement to find work. Although some districts cover a large area, a more likely explanation for much of this movement could be to find somewhere less expensive to live.

(Brant, 1984, p. 30)

What makes flow mapping in some other subjects simpler than within human society is that most other flows are even more local, and mostly in one direction – vector fields. Travel-to-work flows can be seen as an approximation to this; most people go to work near to where they live and travel in the same direction as other commuters. The extent to which this does not apply is shown above. Internal migration (moving home) takes us much further from the ideal, simpler, situation to visualize. Migration flows can be much longer, but

*Daily Commuting Flows
Between English and
Welsh Wards in 1981.*

*50% of all commuters included
10,319,230 people.*

*Flows of more than 2% of the employed
population at the area of residence are
drawn as thin lines to their workplace,
on a land area map.*

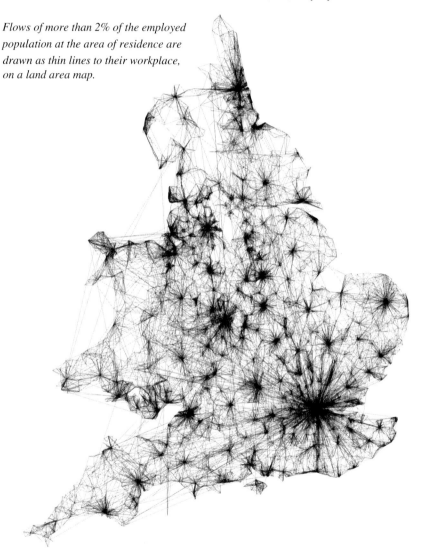

*Figure 6.11   Back on a land map, any flows are now shown when more than
2 % of the employed population of a ward travel to work to the same destination.
The familiar city system of England and Wales is recognisable from just the lines
showing the flows of people to work. The barrier of the Humber River is clear
(the bridge opened June 1981), as is how far some people will travel to work in
London.*

Daily Commuting Flows Between English and Welsh Wards in 1981.

Flows of more than 2% of the employed population at the area of residence are drawn as thin lines to their workplace, on an equal area projection.

50% of all commuters included 10,319,230 people.

Figure 6.12   Population space is seen to be nearly full of the flows of people, and is particularly dense inside London as the cartogram is based on the night-time population. The circular pattern of towns spun around the Capital is distinctive, as are the watersheds between particular conurbations, such as the thin white bands around the West Midlands and Manchester, and also the river Humber.

Daily Commuting Flows Between English and Welsh Wards in 1981.

Where-

All flows which satisfy the following inequality are drawn as thin lines-

$$\frac{m_{ijst}}{p_{is}\, p_{jt}} > \frac{1}{50000}$$

Flows of over 1000 people drawn as thick lines.

m_{ijst} : The number of people moving from place i to j between times s and t.

p_{is} : The number of people at place i time s.

i: Place of residence.

j: Place of work.

s: Nighttime.

t: Daytime.

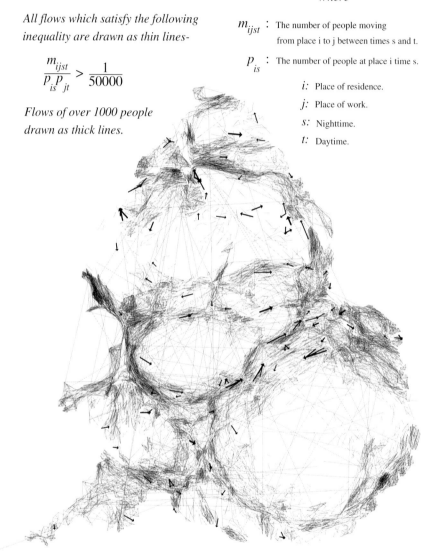

Figure 6.13   Altering the decision over which flows to show can subtly change the picture. This calculation takes account of both the daytime working population and night-time (sleeping) population of the commuters. It shows the rural picture of work flows, but not to large factories, unless, say, over one thousand people come from a single estate, such as one flow to the Cowley car plant in Oxford.

Daily Commuting Flows Between English and Welsh Wards in 1981.

All flows which satisfy the following inequality are drawn as thin lines-

$$\frac{m_{ijst}}{p_{is} p_{js}} > \frac{1}{25000}$$

Flows of over 1000 people drawn as thick lines.

Where-

m_{ijst} : The number of people moving from place i to j between times s and t.

p_{is} : The number of people at place i time s.

$i:$ Place of residence.

$j:$ Place of work.

$s:$ Nighttime.

$t:$ Daytime.

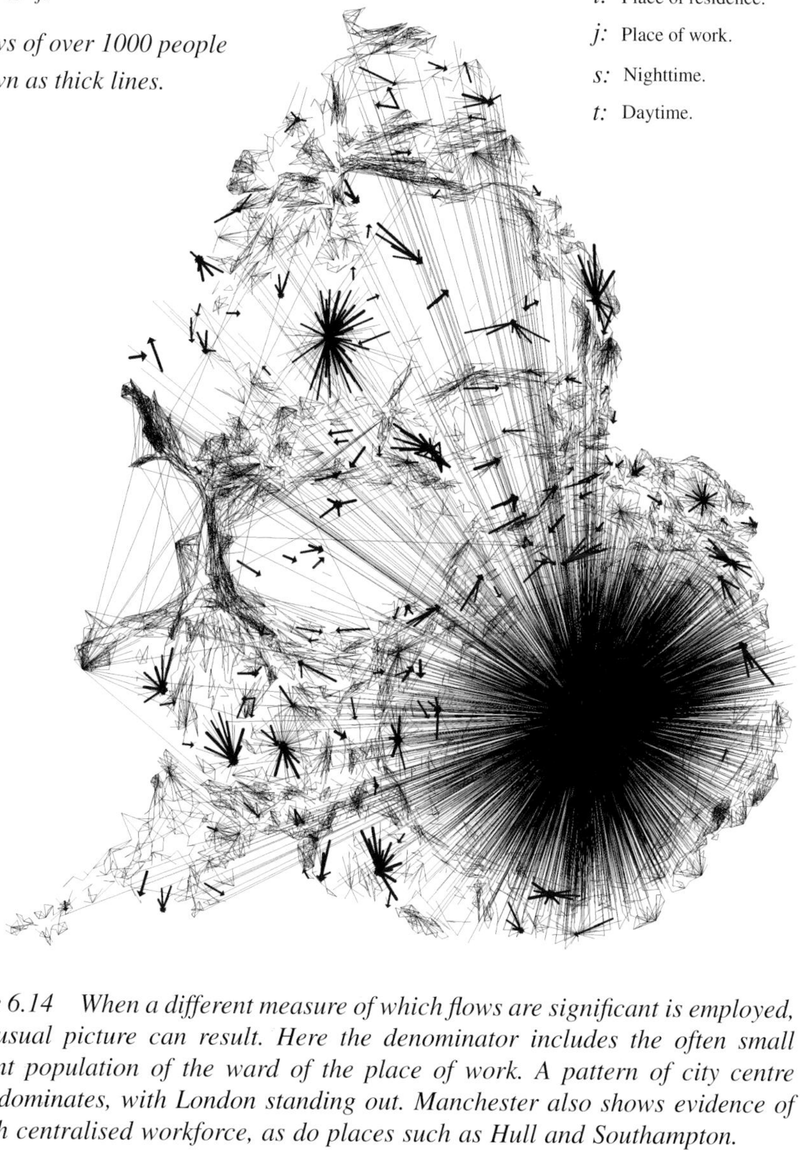

Figure 6.14   When a different measure of which flows are significant is employed, an unusual picture can result. Here the denominator includes the often small resident population of the ward of the place of work. A pattern of city centre influx dominates, with London standing out. Manchester also shows evidence of a much centralised workforce, as do places such as Hull and Southampton.

Figure 6.15   The picture of cities is now augmented by the occupations of those who travel to work in them. The centres of most conurbations are dominated by an orange mix of supervised and intermediate employees, while yellow rivers run down the Welsh Valleys. It is usually the better paid who travel furthest to work. The bright blues of professional enclaves around London are striking.

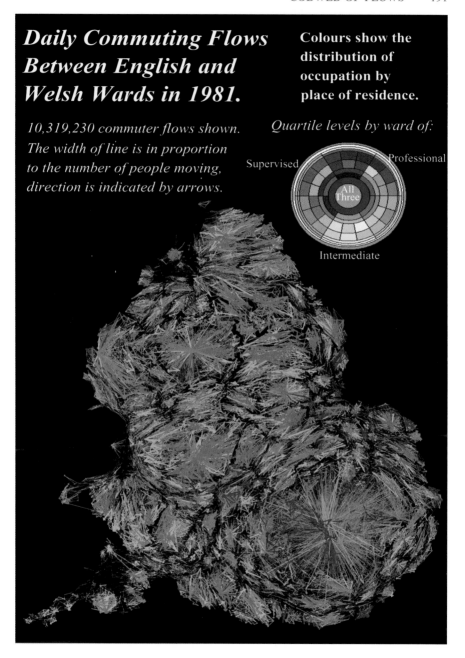

Daily Commuting Flows Between English and Welsh Wards in 1981.

Colours show the distribution of occupation by place of residence.

10,319,230 commuter flows shown. The width of line is in proportion to the number of people moving, direction is indicated by arrows.

Quartile levels by ward of:

Supervised

Professional

All Three

Intermediate

Figure 6.16   The cartogram of commuter flows coloured by social class of commuters is dominated by the orange and yellow areas of the major cities. Some distinctive blue and green streaks run in, and especially around, the Capital. Parallel to some of these flows of people in the best paid jobs are those (red lines) of supervised workers, similarly rushing into this city of extremes.

migration does have a strongly localised tendency.[12] If it did not do so, as we see later, it would be practically impossible to map by the techniques used here (Figure 6.17).

The more serious problem of flow mapping is that there need not be a single strong net direction of flow. Flow can and does occur in both directions. How can bidirectional movement along a line be depicted? Various methods can be used when the flows are only between a few dozen places and the arrows are still large enough to have specific characteristics. Placing the smaller flow as an arrow on top of, and in the opposite direction to, the larger flow is the solution preferred here. Net flow is then represented by the differences between the arrows, as it should be, being a difference and not an entity in itself.[13]

To show net flows only would be simpler, but it would also be highly misleading. The majority of the movements would be largely cancelled out. Interest should primarily be in how the change occurs, in how many people move and where – not in a difference.[14] In fact the net flows in migration are very small in comparison with the absolute movements. Migration flows between two areas tend to be roughly equal in both directions. This fact can be used to advantage when drawing images of migration between many areas.[15]

One solution that has been suggested[16] is to re-route flows by the shortest path through contiguous areas. This was tried, but it was found that for Britain the effect was to imply the opposite of the true picture. Much of the flow went through the middle of the country and the relationship between the metropolitan cities and their hinterlands was reversed. The idea was abandoned.

---

[12] Here we are reaching the limits of what can be sensibly drawn: 'There is no doubt that graphic complexity would be enormous if large populations in sizeable regions for a long time period were to be drawn as paths; the picture of merely one day in a small village is quite complicated even if computer plotters were programmed to do the actual drawing. But the important task of the graphic notation system is not to thrive on visual complexity, but to reveal the under-lying logic of human society and ecology in space and time' (Carlstein, 1982, p. 45).

[13] Some migration counts can include the same person several times, so: 'Although the number of moves into and out of the country during the last ten years was close to 5 million – equivalent to nearly 10 per cent of the total population – this does not mean that 5 million different people were migrating' (Davis and Walker, 1975, p. 2).

[14] '. . . there seems to be a tendency for self-employed and managerial people to move rather less than many of their subordinates, especially over short distances. This might be termed a social effect, reflecting position in the hierarchy of status and power; at longer distances, more economic effects become stronger, so that the poor move much less often than the comparatively rich' (Hollingsworth, 1970, p. 62).

[15] The illustrations drawn here show clearly how the top third of the class structure dominates long distance migration: 'The majority of labour migrants are middle class, in the 25–44 age group, and have middle-high incomes. They are professional and managerial workers in a career structure that encourages movement. Their occupational status frequently entitles them to financial and other forms of aid which make migration easier. They move within a housing market that is especially geared to their requirements' (Johnson, Salt and Wood, 1974, p. 246).

[16] Tobler (1987).

Migration Flows between Family Practitioner Areas on a Population Cartogram in 1976

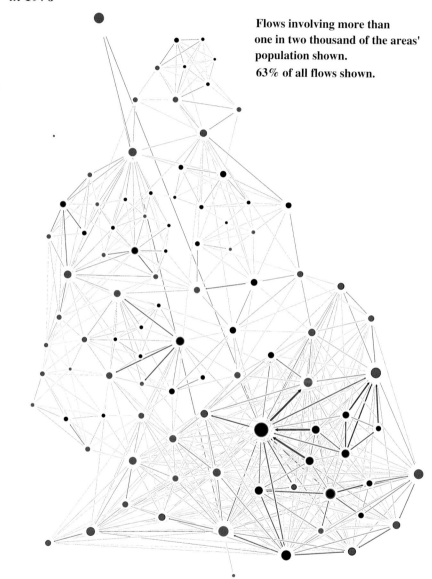

Flows involving more than one in two thousand of the areas' population shown.
63% of all flows shown.

Figure 6.17   Flows between 1975 and 1976 drawn between these 97 areas on an equal population cartogram. At this level almost two-thirds of migrants are represented and Scotland is included, linked most to London (more so than to its neighbours). A division can now be seen running across the country, separating the high propensity to migrate within the South from the much more insular North.

One problem, which has held back work on understanding census migration statistics, has been the method by which flows have been amalgamated when too few people were involved, to a higher spatial resolution. The solution to this problem is similar to that of the changing constituency boundaries earlier – it does not matter. Just as a series of election results can be drawn while the boundaries change, so the lines on the map need not all be between wards but can be between amalgamations, as long as the basic populations of the areas involved are used to determine their significance.[17] In fact, the Census Office may even have done researchers a favour in amalgamating, creating clearer images of the basic movements, which can be coloured according to the attributes of the migrants.

6.7   A space of flows

The process is one of deprived people being left in the urban priority areas as the successful move out to middle Britain.

(Halsey, 1989, p. 22)

Images of migration, the average distance and direction can be drawn, at resolutions as fine as the ward level. These pictures show that migration is a much more diffuse process than commuting.[18] Propensity is most useful; coupled with distance it gives us an idea of how often and how far people move house.[19] Without being able to see the structure of individual streams we are blind to the pattern. We have to guess what there is to see.

Between the ten thousand wards in Britain there could be as many as one hundred million migration streams. In fact only one per cent of these actually occur in a year – one million streams carried some five million people between the years 1980 and 1981. The flows are generally equal in magnitude in both directions. If we amalgamate these two flow propensities as a geometric mean and plot only those that are significantly large, merely one hundred thousand

---

[17] 'Over Great Britain as a whole, the pattern of gross migration levels seems to reflect the political map remarkably well, for Labour Party strongholds correspond quite closely to places where little migration takes place' (Hollingsworth, 1970, p. 33). Contrast this with Hollingsworth's statement above about subordinates moving more. Although appearing contradictory, both statements can and do hold true.

[18] People have been able to travel further to work as car ownership has spread: 'Social changes are likely to add to economic changes in loosening up the pattern of settlement in the industrial regions of Britain in the next generation' (Lawton, 1968, p. 39).

[19] 'Most migrants had moved short distances. In 1981, of those moving within Great Britain about 69 per cent moved less than 10 kilometres (six miles) and only 13 per cent moved more than 80 kilometres (50 miles) – distances measured as the straight-line distance between the grid reference of the address a year before census and the grid reference of the enumeration district of the usual address at census' (Brant, 1984, p. 23).

lines need be drawn, representing the spatial distribution of the movement of the majority of migrants between those years at the finest available resolution.[20]

On the maps shown here each thin line represents, on average, the change of residence of half a dozen families between two wards. Again they can be coloured. This could be done according to the nature both of the area being left and of that being entered, or by any aspect of one or the other. Lines from obviously different areas should stand out from the crowd. As we can see, the pattern of movement is deeply embedded in the social structure (Figures 6.18 and 6.19).

What is striking about the images created is the extent to which migration actually maintains the status quo. The vast majority of movements are within similar areas and the boundaries can be clearly seen. It appears that there is a single, overriding spatial social structure to Britain, a social structure encapsulated by the streams of migration, crossed daily by travel to work. People in Britain must work together or at least very often work within the same places, but they can and do live apart. The patterns of their movements testify to this.[21] These movements are as much part of the spatial structure as are the pictures of who lives in households or works in offices.

With information in the 1980s being released from the National Health Service central register, an opportunity then became available to depict how the flows of migrants were changing over time. A change of flow is a strange concept. If between 1975 and 1976 two thousand people moved from Liverpool to London, and ten years later only one thousand did, what does that reveal? Firstly, one needed to determine whether the overall level of flow throughout the country had fallen, then whether the flow from Liverpool had fallen in general, or that to London dropped, and finally what had happened to the movement in the opposite direction.

Mapping change of flow can involve two images, one showing between which places, and by how much, the relative increases have occurred and the other showing where there have been decreases in the number of moves. Circles centred on the places themselves can show what has happened to their total in, out and net flows. It is difficult to know whether, with the use of contrasting dark and light arrows, these two images could be put on a single map. What certainly

---

[20] 'Migration to and from London is dominant, even when sensible areal units are used to measure it. Migration to and from London dominated population movement in much of Britain: 56 of the 67 largest flows in 1971 involved the London SMLA. There was a tendency for migration to London to come from a rather broader area than migration from London, thus suggesting a process of population redistribution operating through the Capital ... ' (Johnson, 1984, p. 305).

[21] The overall result of all these moves, at a gross scale, was at this time the continued contraction of the major cities: 'The single most impressive finding of the 1981 Census in relation to population distribution was the massive decline in population sustained by Britain's larger cities over the previous decade. The population of Greater London alone fell by almost three-quarters of a million between 1971 and 1981, a drop of almost 1 in 10 .... Even bigger relative rates of decline were recorded by some of the provincial centres, notably Glasgow (-22.0%), Liverpool (-16.4) and Manchester (-17.5)' (Champion, 1989, p. 121).

Yearly Migration Flows Between English and Welsh Wards 1980/1981.

Colours show the distribution of occupation by place of origin.

1,352,520 migration flows shown

The width of line is in proportion to the number of people moving, direction is indicated by arrows.

Equal area projection

Quartile levels by ward of:

Supervised   Professional

All Three

Intermediate

Figure 6.18   Here lines of more than 0.2 % of the geometric mean of the resident populations of the areas of origin and destination are drawn between 9289 1981 census wards on an equal land area map, coloured according to the mix of occupations at the place of origin. The picture tends to be mainly blue as only professionals tend frequently to migrate long distances, but also because of the projection used.

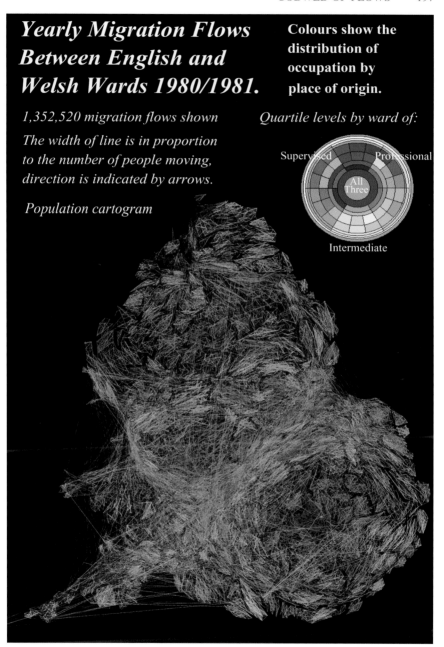

Yearly Migration Flows Between English and Welsh Wards 1980/1981.

Colours show the distribution of occupation by place of origin.

1,352,520 migration flows shown

The width of line is in proportion to the number of people moving, direction is indicated by arrows.

Quartile levels by ward of:

Supervised   Professional

All Three

Intermediate

Population cartogram

Figure 6.19   Locally many moves were constrained by the council house sector to remain in particular bundles of districts or within London boroughs. Surrounding these is a ring of blues and greens, dashed with orange and yellow towns. People move, but they tend very much to move to the same type of estate they left, and so, despite a lot of movement, the spatial class structure is maintained.

could not be shown are the changes over the fifteen consecutive years that were recorded up to 1990 between the ninety-seven mainland family practitioner areas to which NHS migration records were aggregated.

What the images that have accompanied this text illustrate is how the official and conventional place and time based information about people can be turned into pictures depicting the spatial social structure of this country made up from their lives. A tapestry can be woven, the warp and weft of which are made up from the flows of people to work and to new homes, movements in different directions – holding the picture together. Both sets of lines can be placed on a single image, illustrating how the two are linked. Without pictures we would never be able to appreciate such ideas.[22]

Much is missing from this as a view of British society. Flows of money, wealth and goods would be essential to add, as well as knowledge about their static distributions. However, what have been accomplished here are illustrations of ways of extending the depiction of this type of information visually. A change from tables to pictures, from numbers to colours, from words to drawings, allows us to move away from very vague impressions towards more solid images.

---

[22] Migration does not necessarily mean changing who you work for. Even the majority of moves between regions did not involve a change of employer: '... in 1981 over half of inter-regional migrants in the UK (defined as those employed both one year before and at the time of the Labour Force Survey) did not change employer. The figures for 1975 and 1979 were similar' (Salt, 1990, p. 54).

7

On the surface

When I first saw the animation, I watched it over and over again. I thought something like this was going on – but never exactly this.
(La Breque, 1989, p. 527, quoting Ellson viewing a simulation)

7.1   2D vision, 3D world

Many early advocates of visualization claimed the practice began with rendering surfaces. Anything simpler was merely presentation graphics.[1] The thesis this book presents clearly rejects that argument. This was an argument for more expensive machines, rather than arguing for more useful images. The underlying claim here is that, if something can be adequately represented as a two-dimensional image, it is often detrimental to depict it as a more complex object, just as it is better not to use colour unless it is actually needed (Figures 7.1 and 7.2).

Much of today's three-dimensional visualization is unnecessary, and often a damaging embellishment of what is essentially a two-dimensional structure. The primary purpose is usually for dramatic illustrative effect. A dramatic mountain range of unemployment is more interesting to look at than the simple grey shaded cartogram, but is it more informative? We must weigh up the disadvantages of obscuring features, emphasising the foreground, exaggerating the vertical scale,

---

[1] Some claimed visualization should also always be dynamic: 'Visualization of scientific data is very different from graphical analysis or presentation graphics. Visualization implies the use of dynamic graphics to portray changes in an environment over time, or to show the relationships between variables. Dynamic graphics implies rapid update of graphic displays based on operator input, or simulation of real-time changes in an environment through display of movie loops' (Thompson, 1988, p. 1084).

The Visualization of Spatial Social Structure, First Edition. Daniel Dorling.
© 2012 John Wiley & Sons, Ltd. Published 2012 by John Wiley & Sons, Ltd.

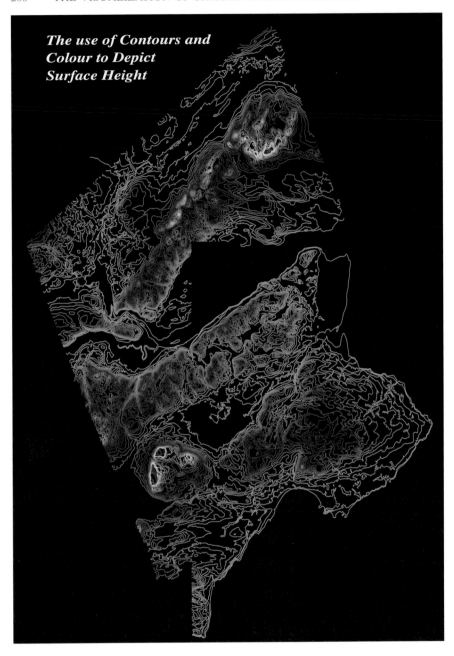

Figure 7.1   Fifty metre contours digitised from Scottish ordinance survey maps on a slightly skewed equal land area projection. The graphic shows how colour can be used to great effect in contour diagrams to show something of the actual shape of complex terrain. Unfortunately, here, some of the hues were darkened in the original printing process, but the overall effect is still dramatic.

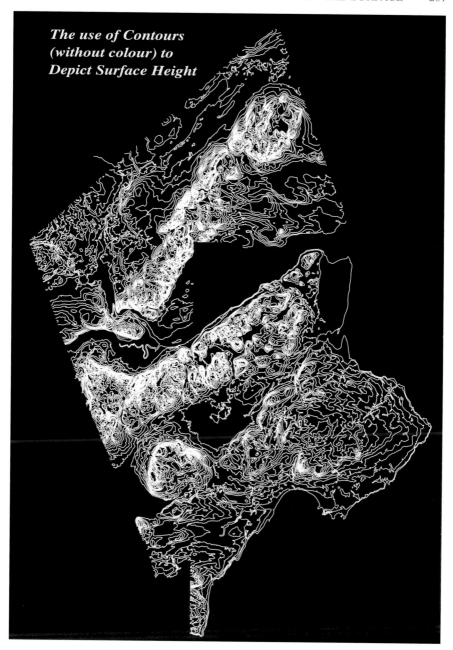

The use of Contours (without colour) to Depict Surface Height

Figure 7.2   Fifty metre contours on a slightly skewed equal land area projection, but now drawn using only black and white. Without some other indication, like colour or the contour heights, we have no idea whether a particular slope is rising or falling. It is also difficult to gauge where the highest points may lie. Contour diagrams do not show the shape of surfaces very well by themselves.

and so on, against the advantage that the eye is used to recover meaning from surfaces. That is what our eyes are most often used for.[2]

We do not tend to think in three dimensions. However, we can become confused when forced to do so. As a result, to give just one example, there is far more two-dimensional than three-dimensional art. It is because surfaces are what we are used to, rather than being implicitly useful, that we consider them here. We have evolved two-dimensional vision with a slight appreciation of depth (much less than many think), but we live in a three-dimensional world. We constantly estimate three-dimensional structure from a series of two-dimensional images. We are quite good at it. The challenge is to harness that ability usefully, to visualize our world through surfaces when it is appropriate to do so, and to do so effectively.

7.2   Surface definition

To recover the lost information from 4D to 3D, we can continuously change the position and orientation of the hyperplane, by either a pure translation or a pure rotation or a combination of both, and obtain different 3D images reflecting all aspects of the 4D. . ..

(Ke and Panduranga, 1990, p. 222)

The surface of an object is the boundary, edge or limit of a shape.[3] It exists in, and encloses, a dimension one order above its own. It contains and defines an object, while expressing form itself. Although the surfaces of concern here are mostly two-dimensional areas in three-dimensional space, it is useful to drop down a dimension to consider another form of surface, one-dimensional

---

[2] Often those producing some of the very best two dimensional visualizations had in their mind's eye something more complex they were attempting to simplify: 'To undertake a project such as the design of this is roughly akin to painting a landscape. One has a mighty scene at one's feet with extensive views and multi-faceted build up. It lives as clouds sweep over it, the light shifts and continuously changing aspects stand out. From all these possibilities of continuously changing pictures the task is to capture precisely that one which is most apposite – for however much the panorama changes before one's eyes, the picture one paints is, even so, static' (Szegö, 1984, p. 17).

[3] 'It has been found that 90 % of people are "3-D blind", including as many as 70 % of engineers working with 3D graphics. The first problem is in design conception. Workers, unaware that they are 3-D blind, are designing components which do not accord with reality. Even top professionals have produced faulty algorithms based on a false 3-D view. Most designers agree with Robin Forrest that "3-D makes life difficult" so structures have tended to be designed in "two and a half" rather than true three dimensions. The second problem is in presenting the 2-D picture of the 3-D artefact. Emphasis has been placed on producing "realism" with a gradually extending set of depth clues: hidden line/surface removal, perspective, shadows, colour and hue, stereo.... We employ enormously expensive systems such as ray tracing to get closer to realism, but if reality itself allows for misinterpretation of the scene, as in all illusions, standard depth clues do not provide a solution and they are not even necessary' (Parslow, 1987, p. 25).

lines, straight and curved, traversing and enclosing two-dimensional space, more commonly referred to as graphs.

Graphs show the relationships between two dimensions. Visually, graphs illustrate the form of the relationship, bring simple equations to life and project complex dependencies. Several graphs can be drawn on a single plane to compare and contrast them (Figure 7.3). Complex graphs can split and merge into many lines, but even a single line can contain infinite complexity.

Much work has been done on how best to present graphs in statistical graphics. The problem begins with the axes. If these are not directly related to each other, the ratio between them is arbitrary and its choice can drastically affect the visual form. Rules have been developed to aid depiction, but, to date, the most successful choice has been to put the graph in a window on the computer screen and allow the viewer to stretch it and view certain parts. This problem reoccurs when we consider surfaces.

Visual improvement in graphing is achieved by transforming the axes more generally. Logarithmic scales are most often used, but anything is possible. Here we have a simple one-dimensional version of the area cartogram problem to solve. A particularly interesting variant is the triangular graph,[4] where the distance of any point from the apexes of an equilateral triangle increases as the influence of what is represented by that apex upon the point declines (Box 7.1). This device is used here to show the share of the vote among three major political parties for a number of areas (Figures 7.4, 7.5 and 7.6) and a little later the shares between four parties.[5] The forms created are extremely interesting given that we now know it was back in the late 1970s and early 1980s that the polarisation that would come to define current British society would begin to grow, hollowing out the middle. Here the beginnings of political separation are made clear, before old Labour capitulated to the schism and 'New' Labour was born.[6]

---

[4] 'The method of using the triangle appears to be one of those things which is continually being rediscovered. The earliest descriptions of the technique that the author has located date from 1964, but it seems likely that others were using the technique earlier' (Upton, 1976, p. 448). See the next footnote for an example.

[5] Use of the triangle's 'third dimension', the distance above its surface, has a long if largely forgotten history: 'Before leaving this subject a brief reference must be made to an ingenious form of solid chart described by Professor Thurston in several of his articles. It is called the tri-axial model. By its use it is possible to take into account four different variables instead of three as was previously the case. It is a necessary condition, however that for each set of corresponding variables three of them should add up to a constant value, generally 100 per cent. The fourth is unrestricted' (Peddle, 1910, p. 109).

[6] The triangular graphs, in hindsight, show key changes in voting patterns marking a new trend in British politics beginning in the 1970s and only hinted at by work done around then. Between the February and October 1974 elections: '... a majority of those in marginal seats who would have [previously] either voted Liberal or abstained if the constituency had not been marginal instead supported the Conservative Party' (Johnston, 1982, p. 51), referring to comments made by Michael Steed in 1975 (Michael later became President of the Liberal Party from 1978 to 1979).

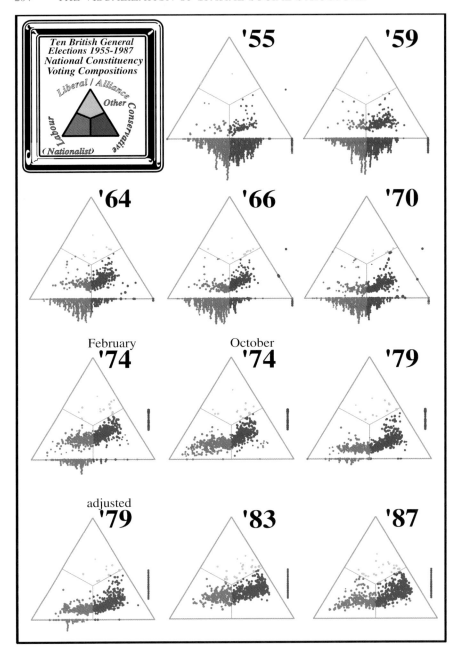

Figure 7.3   Some 705 consistent parliamentary constituencies projected upon eleven electoral triangles, with two party contests drawn as a histogram and Northern Irish dissensions as a line (to the right of the triangles). The change from predominantly two party politics to three is clear, including how the third party fits into the picture. The seats are coloured by the winning party.

Box 7.1   The electoral triangle

An equilateral triangle can show the composition of the votes of three parties, among a number of constituencies, very clearly. Position (x, y) on the triangle is calculated from the Conservative (C), Labour (L) and Liberal/Alliance (A) proportions of the vote as follows:

$$C + L + A = 1$$

$$x = \frac{1 - L + C}{2}, \qquad y = \frac{A\sqrt{3}}{2}$$

The position on the equilateral triangle then gives the share of the votes in any one constituency and the distribution of them all can be seen simultaneously.

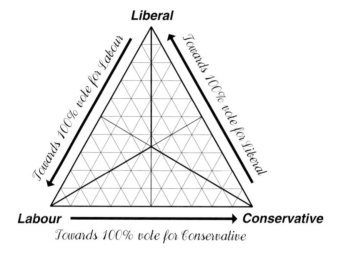

Once the space in which the graph is to lie has been determined, there remains only the relatively simple decision to take on the way in which the data should be drawn. Many different choices can be made. A featureless line is usual, but bar charts and histograms can depict particularly simple cases (rectangles instead of line segments). Scatter plots show the observations upon which the line is based, and can be arranged to show multivariate information. Repeated rendering of convex hulls (or 'kernels') around a set of points produces something akin to a contour diagram.[7]

---

[7] This technique is very similar to Kriging: http://en.wikipedia.org/wiki/Daniel_Gerhardus_Krige.

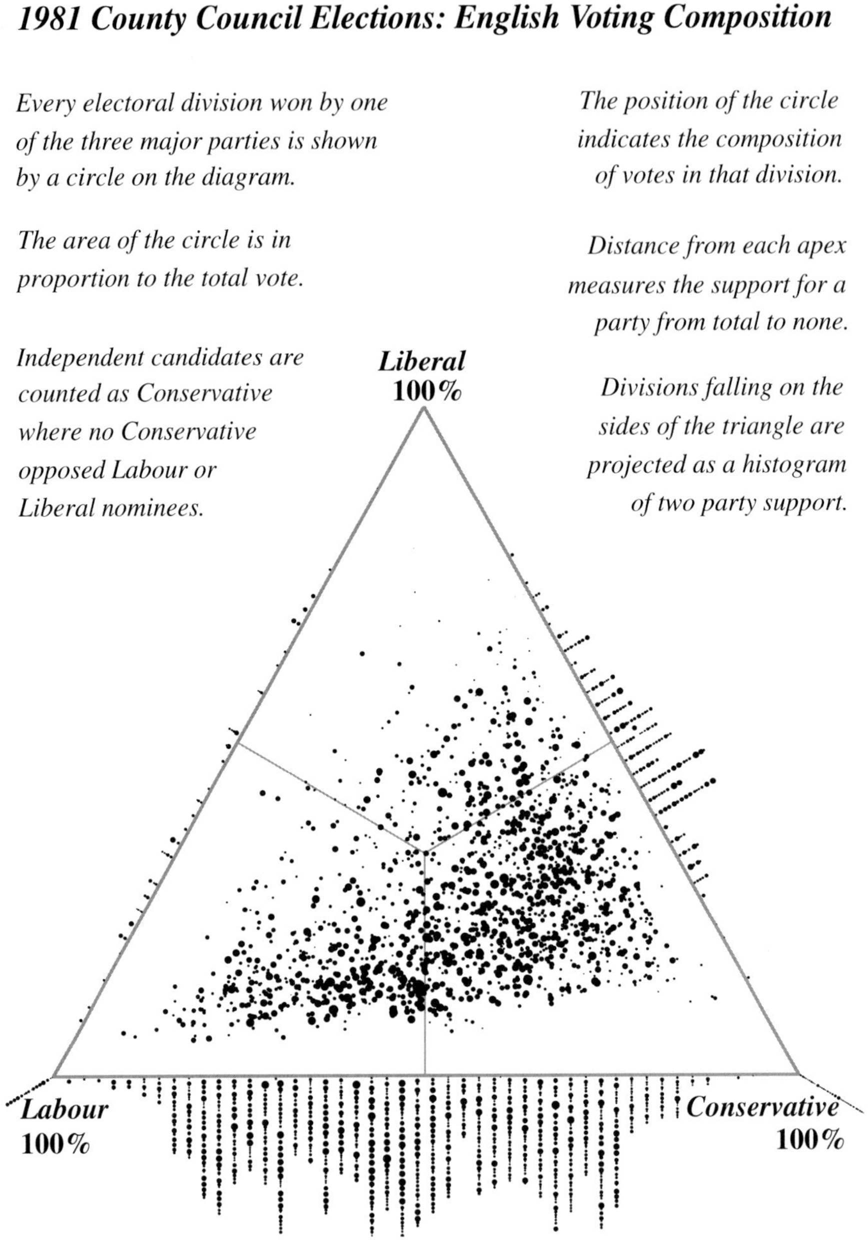
1981 County Council Elections: English Voting Composition
Every electoral division won by one of the three major parties is shown by a circle on the diagram.
The position of the circle indicates the composition of votes in that division.
The area of the circle is in proportion to the total vote.
Distance from each apex measures the support for a party from total to none.
Independent candidates are counted as Conservative where no Conservative opposed Labour or Liberal nominees.
Liberal 100%
Divisions falling on the sides of the triangle are projected as a histogram of two party support.
Labour 100%
Conservative 100%

Figure 7.4 *Around 2855 electoral divisions projected upon the electoral triangle with two-way contests drawn as histograms on the sides and uncontested wins counted as lines at the corners. Circle size is in proportion to the total vote. Local contests involve many more places and greater extremes than general elections. This, and the next few figures, show the results of County Council elections.*

1985 County Council Elections: English Voting Composition

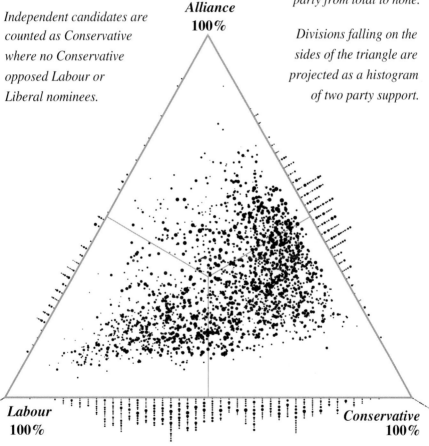

Every electoral division won by one of the three major parties is shown by a circle on the diagram.

The area of the circle is in proportion to the total vote.

Independent candidates are counted as Conservative where no Conservative opposed Labour or Liberal nominees.

The position of the circle indicates the composition of votes in that division.

Distance from each apex measures the support for a party from total to none.

Divisions falling on the sides of the triangle are projected as a histogram of two party support.

Alliance
100%

Labour
100%

Conservative
100%

Figure 7.5   Now 2939 electoral divisions are projected on the electoral triangle with two-way contests drawn as histograms on the sides and uncontested wins as lines on the corners, with circle size in proportion to the total vote. The rise to the Alliance Party, from the situation in 1981, is clear, but a massing away from the Labour Party, up against the Alliance/Conservative axis can also be seen.

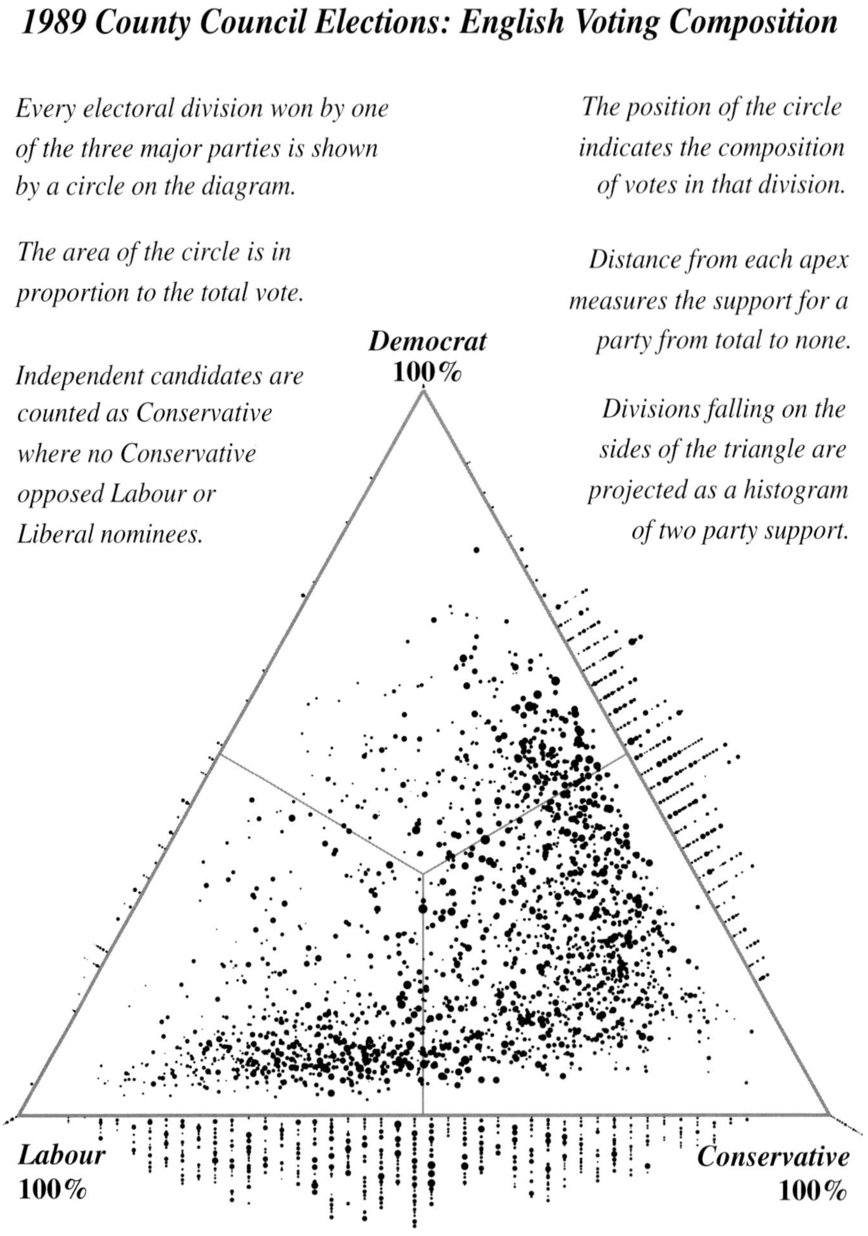

1989 County Council Elections: English Voting Composition

Every electoral division won by one of the three major parties is shown by a circle on the diagram.

The area of the circle is in proportion to the total vote.

Independent candidates are counted as Conservative where no Conservative opposed Labour or Liberal nominees.

The position of the circle indicates the composition of votes in that division.

Distance from each apex measures the support for a party from total to none.

Divisions falling on the sides of the triangle are projected as a histogram of two party support.

Democrat
100%

Labour
100%

Conservative
100%

Figure 7.6   The same 2924 electoral divisions on the electoral triangle. By the end of the decade the sickle shape had become a wedge. The 1980s have resulted in a dramatic political repolarisation of the country. In just the eight years shown here a dramatic change in the political map can be seen, which, even in 1989, was thought might well be reflecting the emergence of a new social order.

7.3   Depth cues

The major problem is that if rotation stops, the 3-D effect disappears.
This is unfortunate because it is helpful to stop rotation to get one's
bearings with respect to the axes; the continuous movement can make
it quite difficult to get these bearings.

(Becker, Cleveland and Weil, 1988, p. 252)

The fundamental problem in visualizing nonflat two-dimensional surfaces is
the need to provide depth cues and their unwanted side effects. These are all the
products of turning what is effectively two-and-a-half-dimensional information
into two-dimensional form: something has to be lost.

The simplest method is to perform an isometric projection of the surface,
mapping all the points in three dimensions on to two dimensions (by matrix multi-
plication). The most basic of these transformations adds half the vertical position
of each point to its horizontal position, then scales the vertical position by half
the square root of three and adds to it the height of the point. Three dimensions
are turned into two and a wire-frame image is produced (Figures 7.7 and 7.8).

To aid perception, a hierarchy of more sophisticated techniques can be
employed. The first of these is to use a perspective projection. Objects further
from the viewer appear smaller (Box 7.2). This obviously distorts the image.
Secondly, hidden lines can be removed so that a wire-frame is no longer seen,
but a more natural solid object is put in its place. Now, however, part of the
original object is obscured. A fishnet of parallel lines can be placed over the
surface, their convergence signifying distance.[8]

More sophisticated options make the image appear more natural. Lighting
the surface from a particular direction creates shadows and more subtle cues, but
lighting distorts any other colouring being used. Ray-tracing makes the surface
even more realistic, allowing for reflections, or more usefully transparency, but
still takes us further from the original form.

The most useful depth cues are to be found in animation, particularly where
the viewer interactively chooses the direction to view from. Rotation of the object,
even simple rocking, helps greatly, although diving with a camera down across
the surface is more dramatic. Parallax is the property being exploited here – the
apparent displacement of objects as the point of observation changes.

All we are doing when we render more sophistication is to make the image
appear more and more like the real world that we are so good at observing. Animation and ray-tracing can be combined to produce stunning images. However,
the difficulty then is in gauging how much of the picture seen is a product of the
techniques required to make it look three-dimensional.[9]

---

[8] There are means of seeing the effect of depth without animation: 'Stereo vision enhances the
three-dimensional effect of the rotating cloud but, even more importantly, the three-dimensional
effect remains even when the motion stops' (Becker, Cleveland and Wilks, 1988, p. 30).

[9] It also remains doubtful how useful stereo vision really is: 'From the test results it can be
learned that for the combined Spatial Map Images the response time is significantly shorter for the

Box 7.2 The perspective projection

The orthographic projection on to image space (u, v) of a point (x, y, z) with the viewpoint at an angle (θ, φ) is

$$u = x \cos \theta - z \sin \theta$$
$$v = x \sin \theta \sin \phi + y \cos \phi + z \cos \theta \sin \phi$$

The perspective projection at a distance (r) and with a particular focal length (f) is given by

$$u = \frac{f(x \cos \theta - z \sin \theta)}{r - (x \sin \theta \sin \varphi - y \sin \varphi + z \cos \theta \cos \varphi)}$$
$$v = \frac{f(x \sin \theta \sin \varphi + y \cos \varphi + z \cos \theta \sin \varphi)}{r - (x \sin \theta \sin \varphi - y \sin \varphi + z \cos \theta \cos \varphi)}$$

For derivation, extension and a full discussion see Plantinga (1988).

7.4 Landscape painting

Applying 2-dimensional tools to 3-dimensional problems has been only moderately successful at best. As the new 3-dimensional geoprocessing tools get into the hands of the users, answers will be discovered to the questions that we currently don't understand or even realize we can ask.

(Smith and Paradis, 1989, pp. 153–154)

As should be realised from the difficulty of visualizing nonflat two-dimensional surfaces on paper, the variability of their structure can be nowhere near as great as that of graphs. Only the simplest surfaces are susceptible to the depth cue method. Most surfaces in our real world are of this simple form.

What is seen in an image containing simple surfaces is not truly three-dimensional, but something just beyond the plane, just the outsides.[10] To visualize true three-dimensional complexity we would have to be able to unravel a ball of wool in our mind, to see all facets and aspects of an object at once, to understand how features would intersect from all around, above and below, and

---

stereo maps compared with the mono maps. However the quality of the answers to the "stereo-questions" does not differ significantly from the "mono-questions". Viewing a Spatial Map Image in stereo means a faster, but not necessarily better, understanding of the map' (Kraak, 1989, p. 112).

[10] Surfaces show two-dimensional elevation, not three-dimensional structure: 'The definition of three dimensional mapping has been incorrectly pre-empted in many cases, by the advertising of so-called 3-D computer programs and video displays that are nothing more than 2-D representations of perspective or similar type projections' (Hardy, 1988).

1981 County Council Elections:
English Voting Composition Surface.

Surface based on the
three party triangle,
height proportional to
numbers of votes.

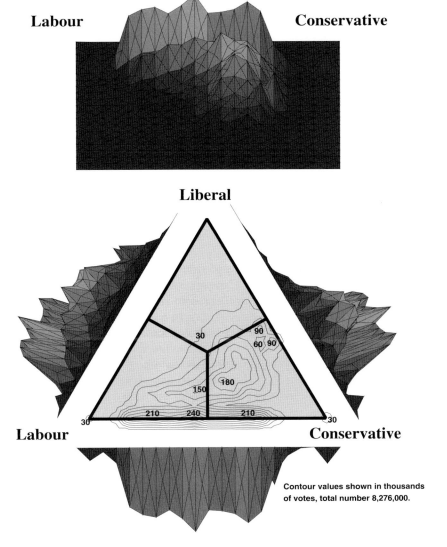

Figure 7.7   The diagrams show how densely different parts of the triangle were populated. Three peaks can be seen along the major two-party axis – each of almost a quarter of a million votes. Then a peak of 180 000 votes is next highest, found in Conservative territory by the Labour margin. The top diagram shows the view, looking down on the surface, from just above the Liberal apex.

1985 County Council Elections: English Voting Composition Surface.

Surface based on the three party triangle, height proportional to numbers of votes.

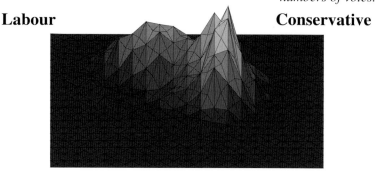

Labour   **Conservative**

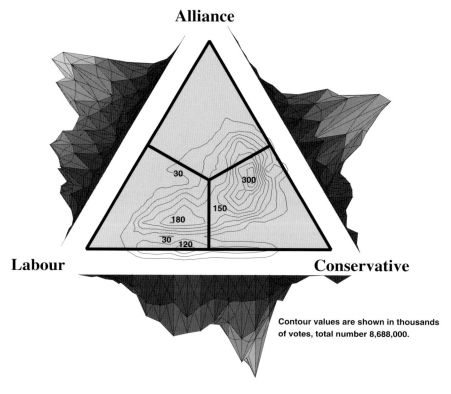

Alliance

Labour   **Conservative**

Contour values are shown in thousands of votes, total number 8,688,000.

Figure 7.8   The surface of local voting changed dramatically in just a few years. The major summit of 300 000 voters was, by 1985, to be found in Conservative territory towards the Liberal border. A second minor peak of 180 000 can be seen here to have formed in the Labour territory as the vote moved away from the Conservative/Labour contests to include the Alliance in the political equation.

to grasp instantly what would result from the rotation of any element in any direction or pair of directions (Figures 7.9 and 7.10). Surfaces do not show us three dimensions; they just persuade us to begin to imagine them. Only a part of visualization is what we see; the other part is what we think.

A major advantage claimed of surfaces is that once one variable is projected as height, other related variables can be shown, say, as surface colour, contours or whatever. This method certainly has its merits. It allows two spatial distributions to be compared before using colour and it dramatically highlights the differences and distinctions (Figures 7.11 and 7.12).

However, in projecting one distribution as shading upon another as height, information is lost and confused. It is lost because it cannot be seen (if it is 'behind') and it is lost as our ability to see and compare difference in (illusory) height is not as good as it is in estimating shades of intensity.[11] It is confused because colour and shadow are created from the projection used and because the shading of the second variable creates the illusion of changes in the height of the first.

Surface shading is not a good substitute for two-colour mapping. The idea of showing the relationship between four spatial distributions by colouring a surface with a trivariate map of colour could only work if the underlying surface were very simple. Where one variable is of great importance and has a relatively simple spatial structure, surface shading of it can be useful.

A simple surface of, for instance, unemployment (Figure 7.13) can be coloured by levels of voting for various parties. Major migration streams could be draped over this, as people, perhaps, flow down from around the mountains of discontent. To create the idea of an industrial landscape this type of depiction can be very useful. However, used like this, it is closer to illustration than visualization – something to present, rather than study.

7.5   Surface geometry

In particular the refutation of Wardrop's conjecture precludes the possibility of constructing a flat map of a city which correctly represents travel time. However, since Warntz's conjecture is true we can construct a curved surface which represents travel time.

(Angel and Hyman, 1976, p. 44)

There is value in using surfaces beyond their illustrative purposes and natural appeal. A surface contains much more information than the mere height measurement, the single variable that is normally extracted from it, and is used in most '3D' socioeconomic graphics.

---

[11] It is claimed that some perspective views are only useful for illustration: 'Traditional methods of representing relief such as hachures, contours, hypsometric tints or hill shading, were developed for topographic mapping and when applied to special purpose maps or thematic maps their effectiveness is often limited' (Worth, 1978, p. 86).

1981–1985 County Council Elections: Surface Showing the Change in the English Voting Composition.

Surface based on the three party triangle, height proportional to change in number of votes.

Labour   **Conservative**

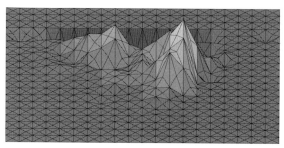

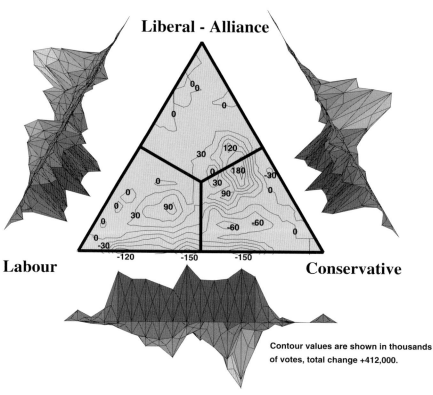

Contour values are shown in thousands
of votes, total change +412,000.

Figure 7.9   Subtracting one surface from another illustrates where most change has been taking place. The two-major-party axis becomes a trench and the rise of three-party politics is shown by a rising new mountain range in the voting landscape. The greatest rise is along the Liberal–Alliance/Conservative political border. The biggest falls are along the triangle's old two-party lower base.

1989 County Council Elections: English Voting Composition Surface.

Surface based on the three party triangle, height proportional to numbers of votes.

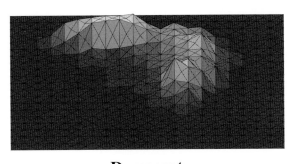

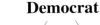

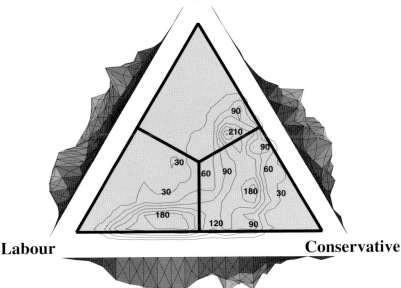

Contour values are shown in thousands of votes, total number 8,295,000.

Figure 7.10   By 1989 the highest peak had, remarkably, moved into the Liberal (Democrat) territory, just over the Conservative border. Other high points were in the centre of the Conservative and Labour territories and, in general, there had been a move away from the centre of the triangle. Places had been sorted out on to a line that curved from Labour to Conservative to Liberal strongholds.

1985–1989 County Council Elections: Surface Showing the Change in the English Voting Composition.

Surface based on the three party triangle, height proportional to change in number of votes.

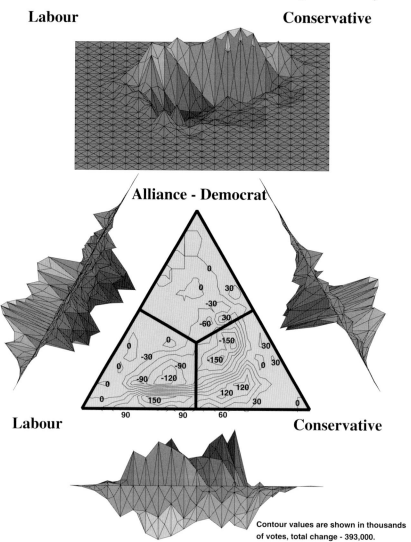

Labour   **Conservative**

Alliance - Democrat

Labour   **Conservative**

Contour values are shown in thousands
of votes, total change - 393,000.

Figure 7.11   The flight from the centre of the triangle is clear as the surface drops back slightly away from the Liberals, but in some places, especially in divisions they already hold, they are still gaining ground. Polarisation is apparent. A growing voting peak of 150 000 extra voters is forming well within Labour territory; two others of 120 000 voters each are now seen in the Tory corner.

1981–1985–1989 County Council Elections: Surface Showing the Change in the Change of the English Voting Composition.

Surface based on the three party triangle, height proportional to change in change in number of votes.

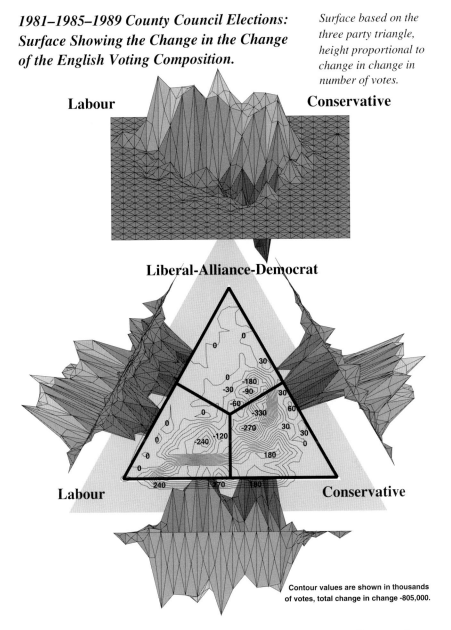

Contour values are shown in thousands
of votes, total change in change -805,000.

Figure 7.12   Overall the second derivative of change shows a collapse in the centre and rise to the edges – much more than just an increase in tactical voting. Gone are the days of compromise politics at the local level. The country by 1989 was clearly divided (for decades to come) between the smallest of places, those localities on the Labour–Conservative axis against those on the Conservative–Liberal line.

The Distribution of Unemployment in Britain 1981.

Rate shown as surface height, volume in proportion to the number of unemployed upon the population cartogram. Resolution 300 by 450, based on ward figures.

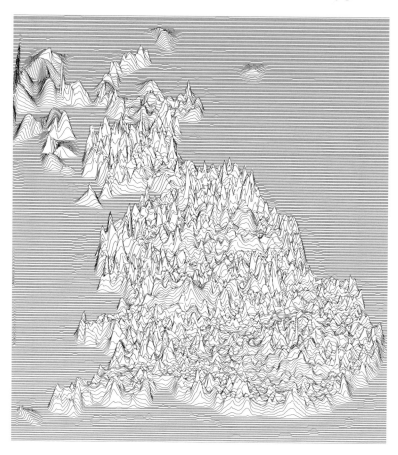

Figure 7.13   A lattice of 300 by 450 points was draped over the distribution of unemployment measured across 10 444 census wards projected on an equal population cartogram. The volume is proportional to the numbers of people out of work, the height, to the proportions. A tall spike can be made out – rising from the centre of the Capital. A ring of low rates can also be seen around London's inner city.

Surfaces define distances between the objects on them. Surfaces can contain spatial information far more complex than that which can be shown on any flat plane. It is this property of surfaces, the geometry they create, that holds most promise for advanced visualization and that has been least exploited to date.[12]

A Euclidean plane (one with 'normal' geometry) has to obey the triangle inequality, which states that the distance from one place to another must be less than or equal to the distance of a route via another location. Euclidean planes are flat; the shortest routes in one are found by following straight lines. On a nonflat surface, however, the straight line distance between two points may well not be the shortest. It is often advisable to travel via another route, such as round mountains, avoiding gorges.

Given a set of distances defined between points, to visualize the space those distances create a surface is formed on which the shortest routes between points are given from a matrix of distances. This matrix has to be symmetrical (the distance is equal irrespective of the direction travelled) and only the shortest possible routes are successfully depicted. Nevertheless, in this surface we have a valuable visual image, which is not a mere elaboration of some simpler information.[13]

Such a surface creates a two-dimensional space in three dimensions, which cannot be arbitrarily stretched and remain valid, although it can be rotated and internally reflected. This property could be used to indicate if real distance were greater in one direction than another, by deciding which way to make uphill and hence which downhill. However, it is uncertain whether this could always be truly depicted and if the ratio of the differences in direction could be shown in a very reliable way.

One further detail of this approach is that the surface could be built upon any two-dimensional, flat spatial distribution. Therefore, when viewed from directly overhead, a familiar geographical picture would be seen, while bringing the orientation of the camera down would show discrepancies from the simpler metric. The most useful possible employment of the technique here is in the depiction of travel time.

---

[12] This means least exploited by 1991 when these words were first written and still by 2011 when they are being edited. Computers are now easier to use and it is easier to print and save files, but much of the early enthusiasm for developing new types of visualization went when the software packages came in.

[13] Tobler saw surface geometry as being of paramount importance in geography: 'The geometry with which we must deal is rarely Euclidean, and it is, in general, not possible to obtain completely isometric transformations.... The maps at first may appear strange, but this is only because we have a strong bias towards more traditional diagrams of our surroundings and we tend to regard conventional maps as being realistic or correct' (Tobler, 1961, p. 164).

7.6   Travel time surface

> *Let us suppose that after an appropriate rotation two dimensions rep-resent the classical longitude and latitude forming a 'basic' plane, and the third dimension, the altitude above the plane thus defined, repre-sents the 'inaccessibility' of a city. The higher above the basic plane, the worse a city's linkages with the global network.*
>
> *(Marchand, 1973, p. 519)*

Geographers have attempted to depict travel time on maps for many years. Because they have usually limited themselves to flat two-dimensional representa-tions, this has proved to be impossible. Correct travel times from a single origin can be drawn, and have been on many occasions. These linear cartograms are created by showing isolines of equal time distance from a point and then trans-forming those lines into circles around that point. Where the travel time space is inverted, however, even depiction of a single point may not be possible in Euclidean space. Imagine what happens as the isolines reach round the globe.

Statistical multidimensional scaling has often been used to try and find the best fitting two-dimensional representation of a set of distances. Frequently all this achieves, geographically, is the reconstruction of the original map with a bit of distortion – only useful when you did not know the original. The essential problem is that travel time, unless exactly equal to physical distance, cannot be drawn on a flat plane,[14] just as, over large areas of the globe, conventional maps distort shape.

The answer to how to create a time surface is to begin with the simple flat geography, and raise or lower points in some third dimension until the correct distances are achieved, creating a surface where distance is drawn in inverse proportion to speed. If you can travel quickly between two points they are drawn close together (at a similar height); if travel between them is only possible slowly then they are drawn far apart by one being drawn much higher than the other.

Just as an infinite number of area cartograms can be created to any given specification, so too can an infinite number of travel time surfaces. The actual algorithm required must create the simplest such surface, containing the least rucks or changes in vertical direction. Thus, for any given Euclidean space, a unique travel time surface can be projected above and below it.

For Britain making a time surface would create a landscape dominated by mountainous inner cities, with London supreme, as by road it takes the longest time to travel into. The major motorways would cut great gorges through the hills

---

[14] A time surface can be defined as: 'Given a velocity field on the Euclidean plane, we define a transformation of the plane into a two-dimensional curved surface lying in three-dimensional Euclidean space. The surface characterized by the transformation has the property that travel time on any path in the original Euclidean plane is equal to the length of the image of that path on the transformed surface. In particular, the image of the minimum-time path between two points on the plane is the geodesic curve joining their image points on the surface. This surface has therefore been referred to as the time surface' (Angel and Hyman, 1976, p. 38).

Box 7.3   Travel time surface

In travel-time space, internal airlines would hang like the lines for cable cars between the peaks representing airports. The surface would undulate smoothly in response to the pressure of traffic on the roads and the general quality of the infrastructure. A main line railway would form a ridge along which settlements cluster in the search for access to work in the city, coupled with the desire to sleep away from it. Occasionally, an international airport may create a hole in this fabric, down which travellers can speed to distant locations.

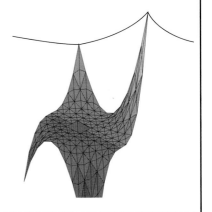

The travel-time surface would show us the economic shape of the country. It may also tell us how some decisions were made as to where to locate factories and why many people live where they do. In some places the surface would be monotonous; elsewhere it could be a tangled mess. It would change with the hours and the years, revealing yet another shape to the country.

of minor roads, or more appropriately tunnels, as they could only be accessed at specific intersections. The ease of access, where it occurred, would need to be made clear. Congested city centres can be cut into by great trunk roads and railways. If internal airlines were included for passenger transport, they might appear as a tightrope connecting the city mountain tops together (Box 7.3).

Euclidean space need not be the two-dimensional basis for such projections upwards. Using Euclidean space in this way only tells of the difference between physical distance and travel time. If a population cartogram were used as the base from which points were projected, so that the distance between points was proportional to the time travelled, the cities would flatten and the land in between rise up. The picture would not be nearly as mountainous as before, as distance in population space is much closer to travel time. The highways and motorways would form a river system into which all other roads appeared to flow, the more minor ones being the headwaters at the highest points on the surface formed.

What is more, upon a population-cartogram-travel-time-surface it would be possible to drape, and see information about, the population between which the roads flow. A multicoloured mosaic of places could be seen rising up in the areas of inaccessibility, spread evenly over the well-connected plains, where the roads were many and the vehicles on them were relatively few. Such an image would help us to understand that the industrial structure of Britain was created using different means of spatial accessibility (coast, river, canal, rail, then road). Such images would be valuable, but no one appears to have yet worked out how to draw one.[15]

7.7   Surface value

That utilization is not simply a matter of physical availability stands out with startling and unfortunate sharpness in Cleveland. The high peaks of hospitals and of physicians are almost literally across the street from the major Black enclave, yet we know the utilization of Blacks to be low.

(Bashshur, Shannon and Metzner, 1970, p. 406)

A surface showing hospital utilization in America illustrates some of the problems caused by assuming smooth continuity – unless very carefully drawn it makes it appear as if those living next to the hospital have easy access. This chapter has shown how surfaces can be created and rendered in as yet largely still to be realised visualizations to depict far more than a series of two-dimensional heights.

Just as a one-dimensional graph shows slope, direction and distance as well as vertical value, a two-dimensional surface can show a multitude of aspects, an entire network of local distances. Other things than travel time or fuel cost could also be shown. Any pertinent variable that can be transformed into a matrix of distances or dissimilarities can be projected as a surface, used to determine the distances between the points on that surface, and then used as a base upon which to conduct further visualization work.

The inverse propensity to commute between wards could be used to show where the divides were strongest, the connections greatest. Social cliffs would appear as real divides, creating between them exposed plateaus and sheltered valleys.[16]

---

[15] The idea of drawing such surfaces was first popularised by Bill Bunge, but he was most concerned about what they would be drawn of, not whether any would be drawn at all: 'Perhaps our almost exclusive concern with such space-warpers is due to the disproportionate influence of economic geography in current theoretical work. We need a grisly "death-miles" distance to explain human migration of a gross planetary sort' (Bunge, 1964, p. 8). As yet not a single surface drawn in proportion to 'death-miles' or even the most parochial of economic variables has been created as far as I am aware, almost half a century after Bunge first publicized them. His inclusion on Richard Ichord's list of people committing so called un-American activities ended his university career. On that list he is placed eight places below Muhammad Ali and one place above Stokely Carmichael. The full list can be found here: http://indiemaps.com/blog/2010/03/wild-bill-bunge/. No other academic geographer achieved as much notoriety. If the 1970 witch-hunt of radicals had not been so effective, apart from much else that could have been better, more progress might have been made in mapping and visualizing societies.

[16] Breaking our thinking out of the plane is an issue that was seen to be of growing importance, especially a quarter of a century ago: 'Even though we navigate daily through a perceptual world of

It has to be remembered that these surfaces can only show the shortest distances between localities. The idea could not be used to show the spatial divisions that long distance migration creates and destroys. What is more, to be successfully interpreted, the surfaces must be relatively simple in form, particularly if they are to form a new undulating two-and-a-half-dimensional, but not incomprehensible, map base.

When the geometry of a surface is not being used, a great deal of compressed visual information is being wasted or, worse still, is misleading the viewer. There are enough valid reasons for using surfaces, without having to use them as a substitute for more simple and effective graphical solutions.

---

three spatial dimensions and reason occasionally about still higher dimensional arenas with mathematical and statistical ease, the world portrayed by our information displays is caught up in the two-dimensional poverty of end-less flatlands of paper and video screen. Escaping this flatland is the major task of envisioning information – for all the interesting worlds (imaginary, human, physical, biological) we seek to understand are inevitably and happily multivariate worlds. Not flatlands' (Tufte, 1988, p. 62).

8

The wood and the trees

How can we display data values representing points in a ten-dimensional data space? What kinds of display techniques demonstrate patterns in such a way that a scientist can perceive those patterns?

(Bergeron and Grinstein, 1989, p. 393)

8.1 Sculptured characters

Consideration up to now in this book has rarely gone beyond the simultaneous visual representations of a handful of variables – three or four at most. Analysts are often presented with situations in which far more aspects are available to be compared than can be compared. What is more, we know that there are usually strong but sometimes subtle relationships lying among all these numbers.

One aim of visualization is to take understanding beyond simple numerical relationships – the idea that when one variable goes up another always goes down. The ideal situation in which to extend understanding beyond correlation is multivariate analysis, where the connections are known to be complex and are usually hardly understood at all. How can visualization illuminate the situation?

The position of an object in the visual plane exhausts our first and most valuable two dimensions.[1] The colour of an object can capture three variables.

---

[1] Some say we can only comprehend four variables simultaneously, some say five: 'At best we may be able to achieve perhaps five dimensions of display using a two-dimensional display plus color. Perhaps stereo displays might achieve six dimensions and animation (time) could in some applications present a seventh dimension. How can we display data values representing points in a ten-dimensional data space? What kinds of display techniques demonstrate patterns in such a way that a scientist can perceive those patterns?' (Bergeron and Grinstein, 1989, p. 393).

The Visualization of Spatial Social Structure, First Edition. Daniel Dorling.
© 2012 John Wiley & Sons, Ltd. Published 2012 by John Wiley & Sons, Ltd.

After employing position and colour we are left with control over the size, shape and orientation of the objects that represent our cases or places. An almost infinite number of subtle alterations could be made to these aspects of the visual representation. It is not the number of variables that can be crammed into its features, but the number of variables that can be visually appreciated and interpreted in each particular context (Box 8.1) that limits and forms our multivariate visualization methods.

The visual objects of concern here are often referred to as glyphs, meaning sculptured characters or symbols (a shortening of the word 'hieroglyphics'[2]). This chapter begins with the simplest of glyphs and moves through to some of the more complex and a little harder to understand, although not necessarily less successful, representations. The aim is to begin to learn what it is that makes glyphs work as visual representations, and how, when and why they fail.

Box 8.1   Areal interpolation

Statistics have often had to be reallo-
cated among areal units in this book.
Where the destination level was a
super-set of the source level, this was
a simple amalgamation. Where the
boundaries of the two did not coincide, the problem was somewhat more
difficult.

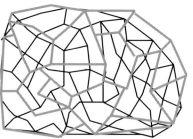

The formulae used to estimate the
value of a statistic (v) from one set of units (i) to another (j) relies upon there
being available a second variable (p) known to be related to the prevalence
of the first variable. The value of the second variable must be known for
every areal unit created from the intersection of the two sets of boundaries
(p_{ij}). The formula is then

$$v_j = \sum_{i=1}^{n} \frac{p_{ij}}{p_i} v_i$$

[2] Others claim as many as nine or more variables can be understood: 'Donna Cox created an innovative technique that clearly displayed a record nine distinct variables simultaneously changing in an animated videotape. To pack variables to such a density, Cox invented a unique 3-D wedge shape, the glyph (from hieroglyphics, the Egyptian pictographs), to represent each computed portion of the flowing plastic. The shape, color (the blue side of the spectrum for pressure and the red for temperature), and orientation of the wedge indicate the state of the flowing material at particular points. The finished videotape shows the plastic (in the form of an army of small wedges) marching into the mould, swivelling, changing direction and color, and eventually settling and hardening in a series of complex steps' (Anderson, 1989, p. 17).

8.2   Circles, pies and rings

This type of presentation makes it easy to grasp the interacting rela-
tionships between age and race. For example, there are tracts in which
most of the children are nonwhite but a majority of the elderly are
white.

(Applied Urbanetics INC, 1971, p. 4)

The basic shape used up to now in this book to represent spatial objects, when they were large enough to have shape, has usually been the simplest – the circle. This choice was made to avoid the shape distracting attention from the overall impression of the image. Rotation has no effect on the circle, but it can reflect one variable through its size.

Size is used here to represent the total population of a place. Circles, when divided into two rings, can be used to show discrete states at two points in time across many places. Could the circle be subdivided further to show the relative sizes of different sections of that population and then coloured by variables such as majority ethnicity at each age?

Pie charts may well be the first possibility to spring to mind. These appear ideal; a dozen subgroups could be shown simultaneously. The circle could, for instance, be cut into male and female slices, these could then each be divided into the proportions working and not working, further subdivided into full-time and part-time workers and so on. There are, however, many problems associated with this.

We are not particularly good at comparing angles, especially when they are presented to us at differing orientations, or at gauging the slight differences in the area of the slices. Worse still, when we are presented with more than a couple of these symbols, we quickly become visually perplexed. We see a multitude of individually complicated parts, which we cannot easily comprehend as a whole.

The basic requirements of a glyph is that not only should it form an acceptable single entity individually but that, when viewed together, a group of glyphs should melt into a gestalt collection so that overall patterns in the multivariate information can be discerned. A group of leaves can combine into differently shaped trees and groups of trees create different looking woods. We must be able to see the woods, not just the trees, from pictures which, without added imagination, simply show the branches and the leaves.[3]

The need for glyphs to work in aggregate as well as alone explains why many initially promising ideas often fail in practice. One method envisaged for

---

[3] Individually well-designed glyphs may fail to combine into a single overall image: 'The dimensions of the trees and castles also lack perceptual integrity. . . . They do not provide their observer a single image or concept or gestalt that he or she can process and remember, binding together the values of all the coordinates of the point. For example, polygons and faces tend to provide observers with such a concept, while glyphs and bar charts tend to look simply like the accretion of their several elements. Trees and castles appear to fall in the latter category' (R.J.K. Jacob, in Kleiner and Hartigan, 1981, p. 271).

depicting the changing spatial distribution of unemployment could be to draw a series of rings inside the circle representing each place. There would be one ring for each of twelve years, like the old bark of a tree trunk, and the rings would be coloured increasingly darker to show the more people there were out of work. It did not work (Figure 8.1). That image, if created, holds little meaning, even when only a few dozen circles are employed. This is because it is not possible to compare across space any more, as space is cut up by circles of time. Each circle is an accurate individual record of unemployment in that place, but the places could now not be seen as a group.[4]

Interestingly, when hexagons are used rather than circles the eye is not so drawn to the centre of each converging set of rings. It is possible to more easily switch between looking at the centres, looking at the outer rings and looking at the middle years. Patterns become more easily picked out, but it is still possible to become overwhelmed by the variety between smaller and larger areas as more urban and more rural parts of Britain diverged in the experiences of unemployment of their respective populations.

One of the most popular forms of glyph in current use is the polygon (often called the 'radar chart'), or its inverse structure, the weather-vane graph. This is formed by representing each case as a point and projecting spokes for each variable associated with that point, out from it at regular intervals, their lengths drawn in proportion to the value of each variable being shown. If the tops of the spokes are connected an irregular polygon is drawn, containing aspects of size, shape and orientation.

This symbol works well when the direction in which the spokes point has some meaning, for instance when showing wind speed in certain directions or the numbers going that way to work. The polygon can produce ambiguous images, however, as two different sets of numbers create the same object. Also, as the number of variables increases, the polygon glyph quickly becomes a formless blob. The basic problem with all of these methods is that they are not producing naturally comprehensible images. What is needed are collections of objects we are used to seeing as a group and for which we already have the visual skills needed to assess as a group.

---

[4] Government in Britain by the early 1980s somewhat reduced the difficulties of visualizing employment trends by reducing the supply of information on employment: 'Between 1971 and 1978 the Census of Employment was held annually and thus became known as the Annual Census of Employment (ACE). In the early 1970s, processing of the data was carried out clerically (which proved costly) but, by the 1977 and 1978 Censuses, computerised processing was underway. However, this was insufficiently planned and led to the delay of the 1977 and 1978 Census results. As a result of these delays and in an attempt to find economies following the Rayner Report, the 1979 and 1980 Employment Censuses were cancelled.... Since 1978, the Census of Employment has been carried out only once every three years, in 1981, 1984 and 1987' (McKee, 1989, p. 9).

The Space/Time Trend of Unemployment in Britain, 1978–1990.

(On a 1981 County population cartogram, outer rings being most recent years - scale indicating deviation from space/time independence.)

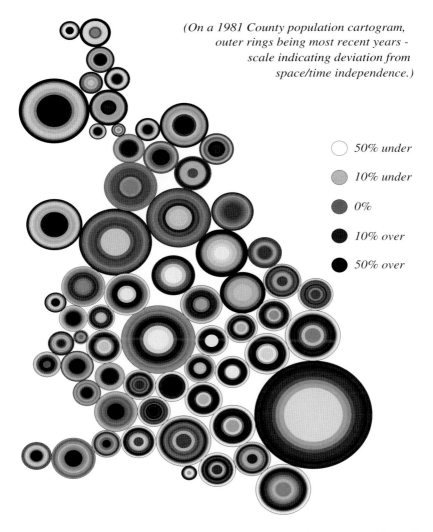

50% under

10% under

0%

10% over

50% over

Figure 8.1   Deviations from expected levels of unemployment are shown here, superimposed as rings around the 64 counties on the equal population cartogram. This method does not work too well. Although some sort of pattern can be discerned, the use of these symbols breaks up the picture and fails to produce an overall impression. The success or failure of an image often depends on these two factors.

8.3   Bars and pyramids

The persistent decline in London's employment over the past twenty-five years or so has occurred despite an industrial structure which has been consistently biased towards activities in which there has been expanding employment nationally.

(Buck et al., 1986, p. 66)

How can we begin to understand how the economy appeared to falter in London and how that was then reversed?[5] If we were trying to show the multivariate information about a single place in isolation, we would probably not draw circles; we would use charts (Figures 8.2, 8.3 and 8.4).

The simplest chart is made up of bars, one bar for each variable, and its height drawn in proportion to the value of that variable. Thus we could show, for instance, the numbers of people employed in eight types of industry simultaneously. As the bars are reflected horizontally and vertically it becomes possible to show the proportions of male and female, full and part time. There are clearly limits to the number of places that could be compared, as the number of aspects we choose to include increases.

One problem with the bar chart is that the order in which variables are placed along the bar greatly influences the visual impression given, and the order is arbitrary. If the order of the industries, say, were made the same as their national ranking, then charts where a gradual rise was broken would show areas where the industrial mix was at odds with what would be naively expected.

The bar chart is taken to one more level of detail when population pyramids are constructed. These are simply two charts placed back to back and standing vertically, usually used to depict the detailed age/sex structure of an area. What is most important about these symbols (as with all glyphs) is that they create a recognisable shape. It is the outline of the pyramid that is important, and this is often simple enough to compare places across space, particularly if differences are exaggerated. Finally, the pyramid can be reflected again, horizontally, to show four related distributions as a cross (Box 8.2), which has been done for some of the illustrations shown here[6] (Figures 8.5 and 8.6).

---

[5] Migration patterns also strongly influence the relationships: 'In the London boroughs the dominance of net out-migration tends to produce different relationships between the components of labour-force change. In boroughs with increasing or stable economically active populations (which are all suburban boroughs), the pattern is generally that net out-migration offsets large increases in female economic activity' (Congdon and Champion, 1989, p. 188).

[6] The role of London is of crucial importance to the developing geography of industrial structure. Some early 1980s writers were especially prescient: 'For what we must remember above all about service activities are that they are growing; that although they are increasingly dispersed within regions, their growth is increasingly concentrated in areas within about 100 miles of London but excluding London itself; and that in this respect especially, and in the close relationship of their distribution to functional areas, their behaviour is unlike that of manufacturing' (Marquand, 1983, p. 134).

Industrial Classification

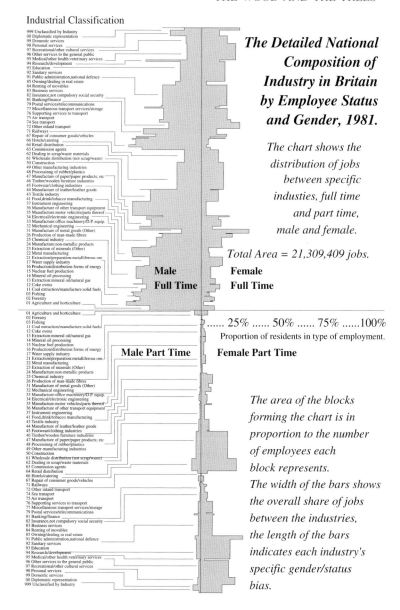

The Detailed National Composition of Industry in Britain by Employee Status and Gender, 1981.

The chart shows the distribution of jobs between specific industies, full time and part time, male and female.

Total Area = 21,309,409 jobs.

...... 25% ...... 50% ...... 75% 100%
Proportion of residents in type of employment.

The area of the blocks forming the chart is in proportion to the number of employees each block represents.

The width of the bars shows the overall share of jobs between the industries, the length of the bars indicates each industry's specific gender/status bias.

Figure 8.2   The mix of employment by employee's status and sex is shown using reflected histograms in over sixty industries for all of Britain from the data provided by the NOMIS national information system. The detailed division of labour between men and women, in full-time and part-time employment, is shown for all the major industries. Almost as many women worked part-time as full-time in 1981.

Industrial Classification

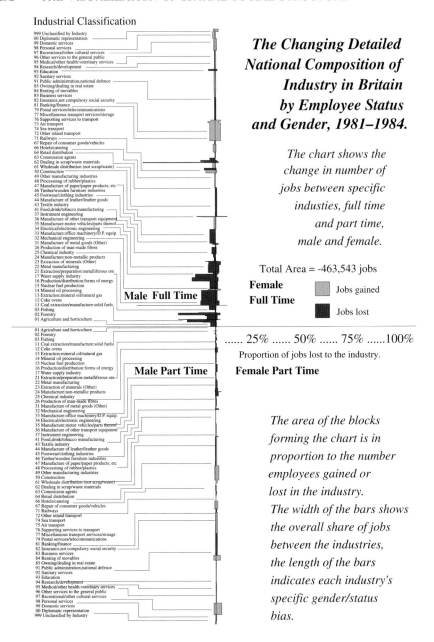

The Changing Detailed National Composition of Industry in Britain by Employee Status and Gender, 1981–1984.

The chart shows the change in number of jobs between specific industies, full time and part time, male and female.

Total Area = -463,543 jobs

Female Full Time

Jobs gained

Jobs lost

...... 25% ...... 50% ...... 75% 100%
Proportion of jobs lost to the industry.

Male Full Time

Male Part Time   **Female Part Time**

The area of the blocks forming the chart is in proportion to the number employees gained or lost in the industry. The width of the bars shows the overall share of jobs between the industries, the length of the bars indicates each industry's specific gender/status bias.

Figure 8.3   Full-time jobs increased in banking and finance, for part-time it was 'personal services', which included jobs like hair-dressing. These increases in work in a few areas were dwarfed by the loss of jobs in other sectors of industry in the early 1980s, most notably from within the manufacturing and mining sectors, usually of the work of male employees and usually of their full-time work.

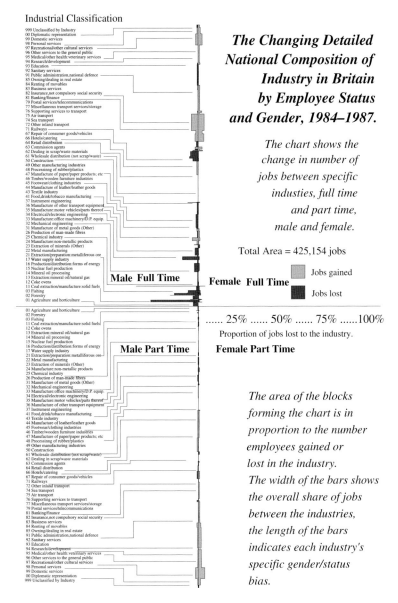

Figure 8.4   The change shows a clear pattern, drawn here to the same scale as the previous image. The loss of male full-time jobs in primary and manufacturing industries is marked, especially in mining related areas. These falls were not offset by the rises (roughly equal for both sexes) in the banking and insurance service industries, especially as so many of the new jobs were part-time jobs.

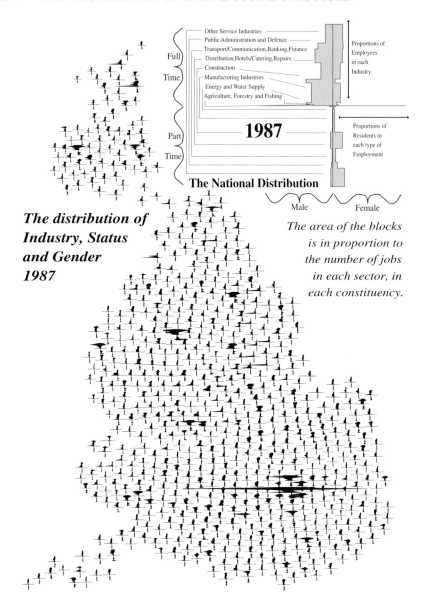

Figure 8.5   Pyramids are drawn, in each of the 633 mainland parliamentary constituencies, showing the industrial structure on the equal population cartogram. The most distinctive feature is the city of London's huge excess of jobs in finance and other services – stretching right across the Home Counties as a result. Many other major city centre anomalies can be found, but none as big.

The National Distribution

Other Service Industries
Public Administration and Defence
Transport/Communication,Banking,Finance
Distribution,Hotels/Catering,Repairs
Construction
Manufacturing Industries
Energy and Water Supply
Agriculture, Forestry and Fishing

Full

Time

Part

Time

Proportions of
Employees
in each
Industry.

0%   Proportions of   20%
Job change in
each type of
Employment

+ 100,000 jobs
- 100,000 jobs

Male   Female

*The Change in
Employment by
Industry, Status
and Gender
1984–1987*

*The area of the blocks
is in proportion to the
number of jobs gained
or lost in each sector
in each constituency.*

*Figure 8.6   Pyramids are drawn, with bars shaded grey for an increase and black
for a decrease, showing the changing industrial structure on the equal population
cartogram. The picture is interesting but very complicated. In the centre of London
the number of full-time jobs for women in services has risen by over 20 % in three
years. Aberdeen South, in contrast, saw the opposite occur.*

Box 8.2   Trees and pyramids

Two unusual glyphs are shown in this book. While appearing very different they share a number of common traits, including multiple variables.

The reflected pyramid is a collection of bar charts showing four closely related distributions. Here, these are of eight industries subdivided by the proportions of male and female, full-time and part-time workers in each place. The area of the symbol is proportional to the number of employees. The height of the bars gives the share of workers in each industry and the width shows how they are spread among the different categories of employment.

A similar use of height and width is employed with the tree glyphs showing house price structure. Here lengths are average prices and widths the number of sales in each sector, giving area as total revenue. Now, however, the combined statistics of sub-markets are shown in branches lower down the tree, the trunk giving the total sales, average price and total revenue for the whole market. The angle at which the branches divide has not been used here, but could be employed to present yet more information.

A fundamental difficulty remains. Bar charts, graphs and pyramids were originally designed to stand alone, and thus often contain enough complexity and detail as single entities, becoming confusing in multiples. Glyphs, to be used in a spatial context, must generalise and simplify the information if the overall patterns are to be understood, particularly if more than a few dozen areas are to be compared.

As the number of areas being visualized grows larger so too do the differences to be seen between those areas. The industrial structure becomes less predictable and the population structure more varied. Unfortunately, at the same time our symbols get smaller and comparison becomes more difficult. We must design simple glyphs that do not require a lot of space and that the eye can quickly comprehend, without excessive examination.

8.4   Flocks of arrows

Overall, despite the decline in the class alignment among individuals, social groups within the British electorate have not become more politically homogeneous. Parliamentary constituencies have never been more politically polarized and, in consequence, the number of marginal constituencies held by small majorities has halved since the 1960s.

(Miller, 1990, p. 49, referring to Curtice and Steed, 1988, p. 354)

Arrows can be used to show far more than geographical flow, as they do (above) in Chapter 6. They are also a glyph that can satisfy many of the criteria of visual simplicity. This is one of the simplest of signs, expressing mainly orientation, although size and shape can also be incorporated to increase information content. Its simplicity allows trajectories at many hundreds of places to be simultaneously shown. Most importantly, at this level, the aggregate begins to express a form of its own – the sum of its parts is greater than those parts as a result of what imagination and intuition add to the overall image.

Like a flock of birds in flight, a group of arrows pointing in a similar direction appears to be going that way; they become a visual group. Arrows can, for example, indicate political flow. This is exactly the impression needed, and it is through an analogy with a natural image that it becomes possible to do this.

Arrows have been used in many ways in this book. The direction of the arrow can represent the levels of two variables as a vector. Here they have been used to show the three-party swing in constituencies between general elections.[7] The arrow can point in the direction that a dot representing the constituency would move on the electoral triangle. The length of the arrow can be used to show another variable – the size of the swing. The size of the arrow is in proportion to the electorate and its colour shows the proportions of the vote going to the three major parties. The position of the arrow is dictated by the constituency cartogram, which could be animated to show changes over time.

In one sense, nine dimensions were being seen in this relatively simple picture – two for position, two for direction, three for colour and one for each of length and size; but that would be a gross exaggeration. The position of the constituency is shown by two dimensions, while the image is representing seven very closely knit variables (only six are independent as direction is one-dimensional). It is the strength of the relationships between the variables that allows so much to be depicted. Ten elections' worth of results can be shown on an A4 page containing over six thousand visible arrows (shown earlier).

---

[7] The same electoral swing does not necessarily imply the same political behaviour in different constituencies: 'In fact a uniform swing could only come about if a party's voters behaved differently, not the same, according to the constituency in which they lived: a uniform 5 % swing from Labour to Conservative logically requires Labour voters to defect at higher rates in hopeless seats than safe seats. That this tended to happen reflected the "partisan neighbourhood" effect' (Crewe, 1988, p. 5).

The arrows worked well in this example because direction was meaningful; left and right as shown here are political terms, not compass directions. They also work well because the spatial relationships in voting were strong enough for discernible patterns to exist.[8] If the purpose had been to look at the effect of changing employment, migration, housing and industrial influences upon the elections visually, these simple arrows would not have been so useful.

8.5 Trees and castles

> ...until Britain moves decisively towards a more-equal society again, its inequalities will continue to express themselves as a north–south divide.
>
> (Lewis and Townsend, 1989, p. 19)

More complex glyphs than arrows have been specifically designed to allow a quick comparison of the overall pattern of multivariate information. The most accepted of these usually take the form of trees or castles, where various aspects of a basic shape are altered to produce many variations of an underlying structure that aids their comparison. It is the maintenance of this basic structure that easily assimilates into a picture that distinguishes these glyphs from the polygons, bars and pyramids described earlier. They have specifically been designed to be glyphs.

Castles have various parapets, which alter in height and aspect as the values of the variables change. In many ways they are simply an embellishment of the bar chart, altered so as to allow the mind to form an impression of the general shape of the place more easily, using a more familiar symbol. Bar charts can only go up or down; they can have a peak here or there, but they are still showing a pattern within the chart. Castles appear more as single objects, and so it is hoped that an overall image can be obtained.

Instead of castles a more familiar alternative of houses could be employed, where the shape of the roof, size of the windows and so on would be altered to show information. Thus a town of houses would be created, allowing particular suburbs, estates and streets to be identified. This method might be especially appropriate if it was aspects of housing amenity at different places that were of interest.[9] House prices for broad categories of housing are shown through

---

[8] The geographical pattern to changes in political swings is not simple: 'Using entropy-maximising estimates of the flow-of-the-vote matrix for each constituency in the 1979–1983 and 1983–1987 inter-electoral periods, this paper... indicates clear geographical variations that are more complex than the simple north–south and urban–rural dichotomies often applied' (Johnston and Pattie, 1988, abstract, p. 179).

[9] It is interesting that places with extreme (high and low) house prices also share the extreme positions in analysing their census data: 'There were six clusters with fewer than five districts including two in which single districts are so distinctive that they each form a cluster on their own.

glyphs in this work using the branches of trees to show the shape as well as local buoyancy of the owner-occupied housing market.[10]

Just as castles grew out of bar charts, trees have grown out of weather vanes. Rather than order the spokes as a wheel, they become the branches of a tree. This works because we are used to seeing trees that vary in their shape but have a rough symmetry about them. The order in which the variables are assigned to the trunk, branches and twigs is crucial for the impression gained. What is usually done is to place the most important variable at the base, and so on up the tree with information of least import altering the lengths and/or angles of the thinnest, uppermost, branches. Whether this works or not depends on the information being depicted.

A relatively convincing wood of glyph trees can be created. Thickets, copses and spinneys of different species can be identified. Overall tendencies for trees to have certain combinations of features in certain parts of the picture, and for other combinations never to occur, can also be noticed on cartograms made up of forests of tree glyphs.

The idea of using the two-dimensional position of the glyphs to show information has often been mentioned, but, because of technical problems and the great caution many cartographers profess to using cartograms, glyphs are rarely used. Glyphs really come into use when the order in which they are drawn on the page has meaning, as well as the order of variables within their own structure having meaning. At this point it really becomes possible to see the wood, through the trees, through the leaves. The images could not be retrieved from their back-up disks to be included in this book at high quality, but showed the 1980s housing division of Britain. They can still be found on the web as low quality scans of Prints CXLVIII and CXLIX of the original thesis.[11]

8.6   Crowds of faces

Both the glyphs and the triangles can raise the dimensionality by two by locating the center on a point in two-dimensional space.

(Chernoff, 1973, p. 365)

The most contentious glyphs created to date are based on human faces, first drawn by Herman Chernoff. Faces, it is argued, are the visual image we are

---

These are the City of Glasgow and the London borough of Kensington and Chelsea' (Webber and Craig, 1978, p. 13).

[10] This market was later to be closely linked to voting. The dependency of voting on other measures of change is a widely held, but infrequently substantiated, hypothesis: 'The Conservative party argued that it was producing a new, prosperous, disciplined country, where enterprise flourished. For those whose local circumstances confirmed that message, there was a greater propensity to vote for that party than was the case for those whose local circumstances indicated that if the government was restoring prosperity, it was doing it elsewhere' (Johnston and Pattie, 1989b, p. 104).

[11] For a link to all the original prints, and more, see: http://www.sasi.group.shef.ac.uk/thesis/prints.html.

Box 8.3   Constructing face glyphs

The face glyphs used in this book are a modern and somewhat less ambitious development of those originally created by Chernoff (1973). Here, only five variables are shown and the faces are made to look somewhat more life-like through the use of curves, rather than lines, to describe them. The faces are each described by a single path made up of eleven Bezier curves, each consisting of two control points and an absolute point (which the curve must lie on). Three curves are used to describe the shape of the face and two each for the eyes, nose and mouth. The control points are shown, shaded in grey, behind the faces.

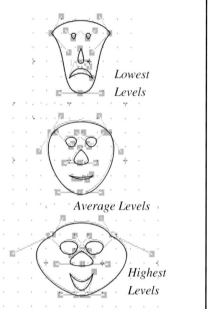

Lowest Levels

Average Levels

Highest Levels

The minimum, maximum and average extent of each curve is shown.

The absolute points remain fixed, ensuring that the general character of the shape does not alter too much and that features will not overlap. The faces used here are symmetrical, as that produced the most pleasing results. After shape, eye size, nose size and smile, the overall size of the face allows up to five variables to be presented at once, in a very novel manner. See code in Appendix to this volume.

best equipped and experienced to decipher. We naturally combine their features to interpret moods – such as happy or sad, or sly. What is more, we can easily compare faces to look for family resemblances or the mood of the crowd.[12] Faces maintain a basic structure in which even slight variation often holds meaning.

The original Chernoff faces aimed to show the values of as many as eighteen variables simultaneously. Here the aim is somewhat less ambitious (Box 8.3 and Figure 8.7). Chernoff faces have been used here to study general election

---

[12] Possibly the first suggestion of drawing crowds, or at least not objecting to such a drawing on a cartogram was made here: '... engineers prefer line graphs, sales people bar charts, demographers pie charts and medical personnel lists of numbers. Epidemiologists, at least those dealing with cancers, seem to appreciate horizontal bars. In cancer statistics and epidemiology the discrepancy between sophisticated statistical methodology and elementary graphical techniques is large. Certainly, elegant technical refinements can be found in cancer mapping, but even here there is exciting potential for maximizing the information content of maps by combining cancer frequency levels with, e.g., indices of data quality. Moreover, no objections exist to combining cartograms and faces' (Rahu, 1989, p. 765).

625 Chernoff Faces Showing All Permutations of 5 levels of four features - cheeks, eyes, nose and mouth.

Figure 8.7   Most of these combinations do not actually appear in the following pictures, so it is entertaining, and perhaps useful, to see them all here. Of all potentially possible different combinations of facial features, what emerges tends, like in real life, to be limited to what goes with what, due to correlations between variables. However, no combination produces an implausible human face.

voting patterns in Britain in relation to just five variables – the electorate (head size), house price (thin face for low, fat for high), employment (smile), election turnout (nose size) and industrial structure (large low eyes for younger industries).[13] The colour of the face can then represent the actual voting patterns, when the faces are arranged as a group on the constituency cartogram (Figure 8.8). It is also well worth looking at the faces with the colour taken out of them (Figure 8.9). The computer code used to draw these faces is reproduced in the Appendix to this book.

The images of smiling and frowning heads were initially intended to be a tongue-in-cheek extension of Chernoff faces, faces amalgamated to the level of crowds. Nevertheless, the inner-city/outer-city and north/south divides in many aspects, as well as voting, can clearly be seen (Figures 8.10 and 8.11). The difficulty of drawing precise lines between the regions and around cities is clear. What is more, specific outliers can be identified, which do not fit in (just as before with arrows that did not go with the flow). Variables that appear to be unrelated to the rest of the picture can be identified. Complex three- or four-way interactions, where certain levels of some variables apparently combine to produce a particular effect, can also be identified.

The use of the population cartogram developed here, as the spatial base for these faces, has particular advantages. None of the faces overlap to obscure each other and they are all, by their size, drawn in proportion to their importance based on the numbers of people they represent. It may well be that glyphs have not been much used in spatial images before, because of the dire problems of spatial congestion, which the use of population cartograms in place of land area maps overcomes. It would be difficult, for instance, to use faces in place of the circles that overlap and cluster on the electoral triangle, although it may produce an interesting picture to compare with the crowd on the cartogram.

Strong local relations in space are perhaps the clearest message formed by the images. Sharp divisions are also immediately apparent, as are more gradual changes.[14] The faces can also be used to show that the changes over time in the variables might be contributory to the changes over time in voting. Thus the expressions become places' reactions to a changing situation, their colours perhaps indicating some of the electoral results of those changes (Figures 8.12 and 8.13).

Chernoff faces are contentious for some of the same reasons that they are so useful. People's reactions to faces are much stronger than to more neutral objects, which are claimed to depict the information more objectively. Facial features are inevitably interpreted as depicting joy or sadness, concern or apathy

---

[13] Research has found that even slight changes in expression are perceived: 'This latter finding suggests that extreme caricature like faces are not crucial in obtaining good performance' (Jacob, Egeth and Bevin, 1976, p.193).

[14] Faces can provide an alternative to the use of aggregate indices in studying multivariate spatial change: 'The over-riding impression of the changes taking place in local economies since 1981 is clearly of the division between north and south' (Champion and Green, 1989, p. 84).

and so have to be carefully assigned to variables. Cartographic preferences move along a continuum, from personal likes and dislikes of certain colours in maps, to individuals' reactions to cartoon faces. Visualization, at a higher level, is all about engaging our imaginations and emotions.

8.7 Information overload

> *Undoubtedly, the faces give a more attractive gestalt impression than the other symbols; people like to look at them.*
>
> *(Kleiner and Hartigan, 1981, p. 261)*

A serious criticism of the use of glyphs is that they can overload the viewer with information. Too much is being asked of the eyes and the mind.[15] In this chapter an illustration has been included of badly designed symbols, rings of unemployment rates, that are impossible to decipher spatially. It has also been suggested that well-thought-out images can help the viewer form higher level structures out of the simple pictures of collections of places. A most efficient way to achieve this is to use arrows, but these require direction to have some meaning (such as political left and right) and can show only a few other related features.

The creation of crowds of faces is certainly the most ambitious use of symbols. This may well be the first time they have been used so extensively in this way – pictures of people's faces on paper, drawn to show information about people in places. In hindsight it is obviously an interesting idea. Cartograms where every place is given the space of a circle make the idea realisable.

Whether these glyphs work spectacularly, or not at all, one thing is for certain: they get people's attention and make them think.[16] That might partly be because viewers are confused by the amount of factors to consider, but it can also be because they understand it, and begin to see so much more. The use of symbols that bear some relation to the subject being studied is an asset. How better to show differences in the sizes and quality of houses than by a collage of images of those houses? How better to show factory closure and growth than with pictures of industries being born and dying?

---

[15] '... maps portraying more than one aspect (variable) of a phenomenon are being published in increasing numbers and ... the comprehension and understanding of these maps is likely related to some basic structural characteristics of the maps' (White, 1984, p. 45).

[16] We are, however, used to seeing and understanding initially confusing situations: 'Under natural conditions, vision has to cope with more than one or two objects at a time. More often than not, the visual field is overcrowded and does not submit to an integrated organization of the whole. In a typical life situation, a person concentrates on some selected areas and items or on some overall features while the structure of the remainder is sketchy and loose. Under such circumstances, shape perception operates partially' (Arnheim, 1970, p. 35).

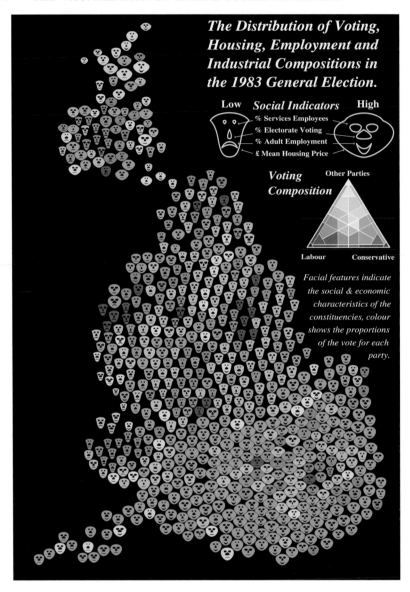

Figure 8.8 The death-like heads inside Glasgow city are solidly orange (Labour/SNP) while the happy faces around the Capital voted solidly for the government of the day. The Welsh may not have many jobs or expensive housing, but they still turn out to vote in large numbers. This technique is particularly good for identifying exceptions, such as Tynemouth, Newham South and the Isle of White – all odd places. Figure 2.2 shown earlier provides a key to the areas.

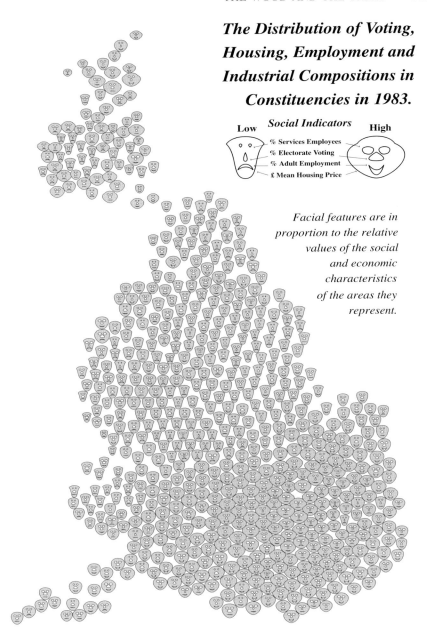

The Distribution of Voting, Housing, Employment and Industrial Compositions in Constituencies in 1983.

Low   Social Indicators   High

% Services Employees
% Electorate Voting
% Adult Employment
£ Mean Housing Price

Facial features are in proportion to the relative values of the social and economic characteristics of the areas they represent.

Figure 8.9   It is preferable to omit voting colour when first seeing how the country looks. The large fat and especially the happy faces in the far north of Scotland are more visible here and London doesn't look quite as jolly within the centre of that city when not distracted by political colour. The emaciation as you move towards the North West is also more apparent. Colour always dominates.

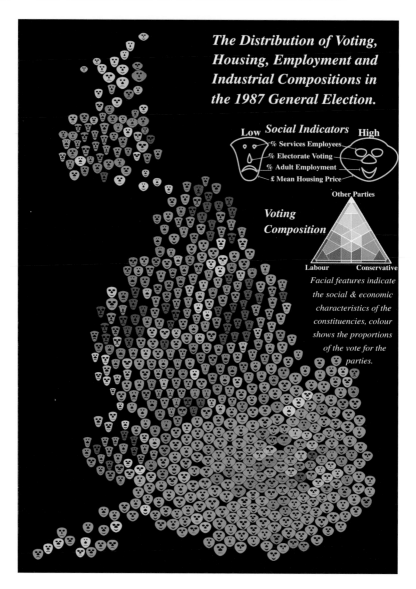

Figure 8.10   The 633 mainland parliamentary constituencies are each repre-sented by a face whose features express the various variables and which are coloured by the mix of voting. Particularly outstanding are the colours of Oxford East, Yeovil, Norwich South, Tweedale and Glasgow Eastwood. Several expressions have changed markedly since 1983, but are harder to find on this constituency cartogram. See Figure 2.2 in Chapter 2 for area labels.

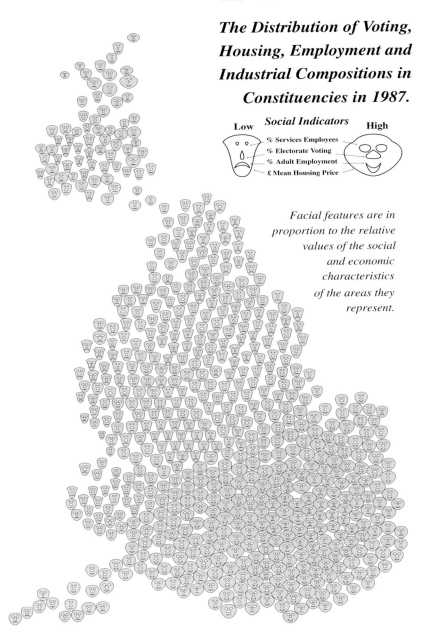

The Distribution of Voting, Housing, Employment and Industrial Compositions in Constituencies in 1987.

Low   *Social Indicators*   High

% Services Employees
% Electorate Voting
% Adult Employment
£ Mean Housing Price

Facial features are in proportion to the relative values of the social and economic characteristics of the areas they represent.

Figure 8.11   There are more smiles within Scotland and Wales, it would seem, when the colour is turned off. The North of England was happier too; at least some jobs were reappearing. The Home Counties appear positively ecstatic. This cartogram can also be compared with the previous black and white one for 1983. The South West of England is more sanguine, as is the east end of London.

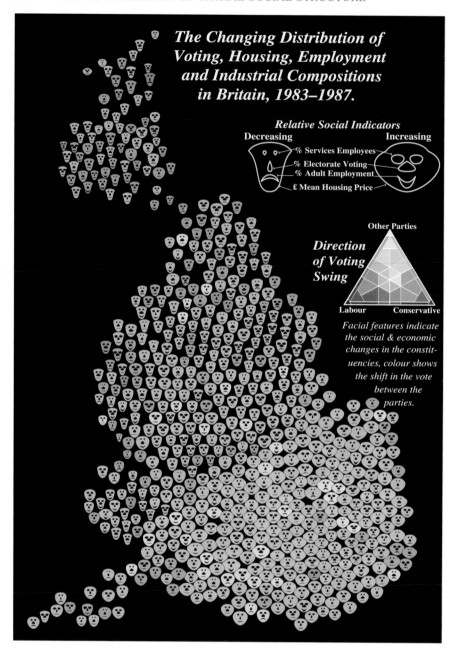

Figure 8.12   Relative shifts in the political vote overlay the other changes. Croydon North West is upset and Taunton is none too happy, while Aberdeen South has become most distressed. Here we can see many things at a glance and quickly estimate their relative importance – is a particular relationship a general tendency or an isolated case? This image is of the colour and reaction to change.

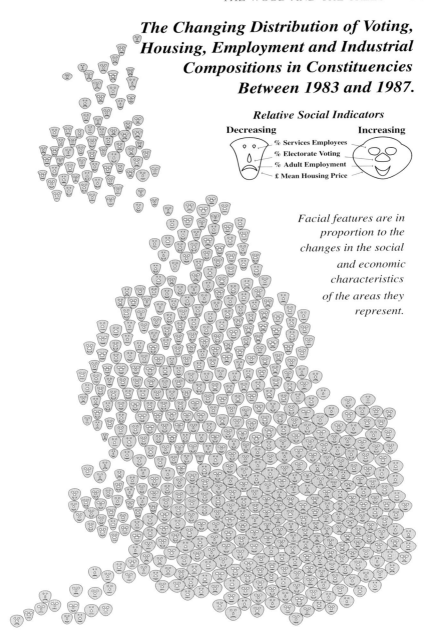

The Changing Distribution of Voting, Housing, Employment and Industrial Compositions in Constituencies Between 1983 and 1987.

Relative Social Indicators

Decreasing   Increasing

% Services Employees
% Electorate Voting
% Adult Employment
£ Mean Housing Price

Facial features are in proportion to the changes in the social and economic characteristics of the areas they represent.

Figure 8.13   Look at the South and south east London; fat but miserable faces. Was this distress caused by the loss of jobs there or gaining few? In contrast it is happy days in the North and West despite shrinking faces. Who cares if your homes are not rising much in value if you have work? In Scotland, south of Aberdeen, more joy is coming. Expensive housing does not imply happiness.

It is a mistake to think that these symbols can add another dimension to the two we have on paper.[17] Glyphs show multivariate structure, not multidimensional form. We can look at a lot of categorical aspects of many places in space simultaneously. We cannot see how some feature varies with, say, place and varying wealth, multiparty voting or disease in spacetime.

Varying the features of an object is not a good substitute for varying its position. Features of an object have no geometry and thus a limited ability to show only a few values. To get a real extra dimension, beyond the first two, we must begin to think in terms of volume and the next, penultimate chapter, of this book.

---

[17] However, the more effective the technique, the greater is the information that can be shown: '... the ability of humans to analyze effectively spatial distributions is alleged to deteriorate progressively as the number of variables increases, inter-relationships among variables becomes subtle, and the magnitude of variations decreases. This suggests that cartographic presentation must demand as little mental computation and conceptualization as possible if the full potential of creative intuition and decision making is to be realized. If the cartographer can develop more effective data reduction techniques, and the map reader can be taught to understand their underlying concept (i.e., readily decode them), then the amount of information communicated by a single map might be greatly increased' (Muehrcke, 1972, pp. 19–25).

9

Volume visualization

Once time becomes a dimension within which activities can be viewed,
the map, because of its static cross-sectional view of phenomena, loses
its usefulness.

(Holly, 1978, p. 12)

9.1 The third dimension

You could be forgiven for thinking that the third dimension had already been
utilised in this book, but that is not so. A dimension is something that can be
both measured and moved around in, allowing the existence of a geometry – the
relative arrangements of objects in space. Thus, real world time is not strictly a
dimension for humans, as we cannot move around in it.[1] As explained before,
but of even greater importance here, although we live in a three-dimensional
world our vision largely waters it down to two dimensions, which our mind then
reconstructs into a particular kind of three dimensions. The world does not appear
to go flat when you shut one eye, but what you think of as the three-dimensional
world is a very human simplification of it.

We mostly think with two-dimensional concepts, and that can be used to
advantage in visualization. Time can be viewed as a third dimension in the
social world when phenomena beyond the simple single lives of individuals
are being considered. A social order of opportunities, jobs, customs and culture
exists and moves in time and space. A disease is a spacetime entity and its social
repercussions can only be understood when it is seen as such, knowing when, as
well as where, it strikes.

---

[1] 'It certainly feels like time is passing; I'd be foolish to argue otherwise. But I want to show you
that this feeling is a sort of illusion. Change is unreal. Nothing is happening. The feeling that time
is passing is just that: a feeling that goes with being a certain sort of spacetime pattern' (Rucker,
1984, p. 140).

The Visualization of Spatial Social Structure, First Edition. Daniel Dorling.
© 2012 John Wiley & Sons, Ltd. Published 2012 by John Wiley & Sons, Ltd.

When does a variable become a dimension? That is essentially a question of the resolution of measurement. If place is just one of the dozen regions and nations of the British state then it is a variable that should be put in tables, not mapped. Once a variable has numerous possible values it can be considered a dimension. Movement and measurement along the variable must be possible, and the three-dimensional space created should theoretically be continuous. Time in the study of the ten 1955–1987 general elections is too discrete to consider as being approximately continuous, and interpolation of votes between the elections would be meaningless.

If we have a third dimension, how can we see it, let alone understand it?[2] This is the problem that is responsible for relegating this chapter to the end of the book. Basically, the answer is – not easily. The traditional way to see in such blocks is to show some two-dimensional slices, as we might cut open a human brain in a medical scan. It is a small step from there to take many slices, allowing animation. To create more of an illusion of three dimensions, perspective and various lighting effects can be employed. These too can be animated.[3]

All that we are really showing with traditional three-dimensional graphics is a series of surfaces – two-and-a-bit-dimensional, but a long way from three – often containing almost one-dimensional information (Figure 9.1). During recent decades many innovations were made in computer visualization that can create far better illusions of more complex three-dimensional worlds. The problem is then no longer deceiving the eyes, they are easily deceived, but teaching the mind. However, in the social sciences it simultaneously became fashionable to move away from looking at any illustrations based on numbers just as all these new ways of looking at numbers were being invented.

9.2   Spaces, times and places

It is by positioning our geography between space and time, and by seeing ourselves as active participants in the historical geography of space and time, that we can, I believe, recover some clear sense of purpose for ourselves, define an arena of serious intellectual debate and inquiry and thereby make major contributions, intellectually and politically, in a deeply troubled world.

(Harvey, 1990, p. 433)

---

[2] We are well equipped for visualization, but still often find it difficult: 'It is estimated that fifty percent of the brain's neurons are involved in vision. 3D displays light up more neurons and thus involve a larger portion of our brains in solving a problem' (van Dreil, 1989, p. 2).

[3] Animation is almost always required to gauge depth correctly: 'Although it is not obvious why it should be, small, rapidly repeated, changes in the viewing transformation are seen as continuous motion of a rigid object – the point cloud. We automatically see the three-dimensional shape of the point cloud, using the unconscious human ability to perceive shape from motion' (McDonald, 1988, p.184, quoting from Marr, 1982).

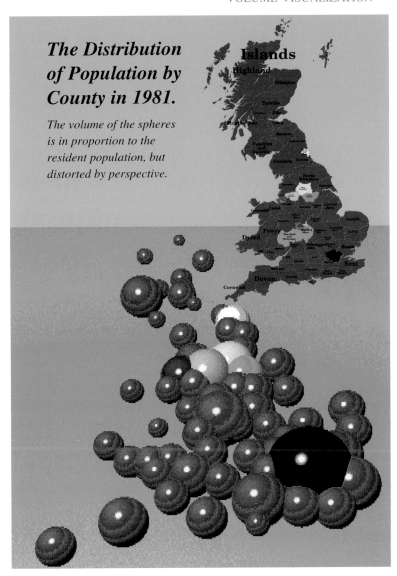

Figure 9.1   The 64 counties are shown as spheres, the volumes of which are drawn in proportion to their populations, coloured to match the colours in the map and projected on an equal land area map, viewed from the south. Three-dimensional views can be misleading. Here, Devon appears larger than Strath-clyde because it is drawn in the foreground, while the size of London obscures other areas. Note: the lower part of this image is an enlarged low resolution computer screen snapshot that was created by early ray-tracing software and so appear mottled in places.

Places exist temporarily as well as spatially. Over the years people move home; over the decades new homes are built and old ones decay; over the centuries towns are formed and decline (Figure 9.2). An animation of the national boundaries of the European continent over the last four hundred years would show near continuous turmoil. Nations exist only as pockets in space and time, as also, in the long run, does the world system of nation states. Regions coalesce, fragment and disappear.

The plate tectonics of human geography is a violent spectacle. Even the patterns of spatial inequality can alter in the space of a few dozen years. We can never appreciate why something is if we do not look at how it came to be and what it is becoming.[4]

The theme of unemployment has arisen at several points in this book. Unemployment as a national phenomenon has a well-defined spacetime geography. It did not exist in Britain, or almost anywhere in the world, as a recognised phenomenon prior to 1888.[5] Ninety years later is had risen, fallen and was just starting to rise again. Monthly records have been kept by the eight hundred and fifty-two amalgamated office areas for every month since 1978 (and for every ward since 1983). Over one hundred and fifty temporal observations constitute a dimension in the above sense. How then to view this information?

In the research work undertaken that resulted in this book many attempts were made to show the spatial structure of unemployment over time; some have been mentioned in the previous chapter. The structure was just too complicated for a few views into a spacetime block to uncover (Figures 9.3 and 9.4), so a series of time-slices were drawn, one for each year since the series began. To highlight the changes, deviations from the expected value were calculated, knowing the average for the year and the place. If this had not been done, the changes over these twelve years would not have been visually apparent. Similar problems were encountered when trying to show house price change in a single image (Figure 9.5).

Twelve cartograms of unemployment were created for this book using both counties and amalgamated office areas to show how spatial resolution changed the image (both are shown earlier, using circles and hexagons respectively). One picture was drawn a year, partly because that is all that would fit on the paper

---

[4] Appreciation of spacetime requires us to take an unfamiliar vantage point: 'My world is, in the last analysis, the sum total of my sensations. Sensations can be most naturally arranged as a pattern in four-dimensional spacetime. My life is a sort of four-dimensional worm embedded in a block universe. To complain that my lifeworm is only (let us say) seventy-two years long is perhaps foolish as it would be to complain that my body is only six feet long. Eternity is right outside of spacetime. Eternity is right now' (Rucker, 1984, p. 136). Perhaps unsurprisingly, Dr Rucker is related to the philosopher Hegel and so far his lifeworm is doing fine, but he is not yet 72; see: http://en.wikipedia.org/wiki/Rudy_Rucker.

[5] Understanding that people did not lack work through their own failure was a huge step forward in the 1880s. Before relief for unemployment began people faced starvation, the workhouse or doing any job no matter how demeaning or low paid (Dorling, 2011, p.15).

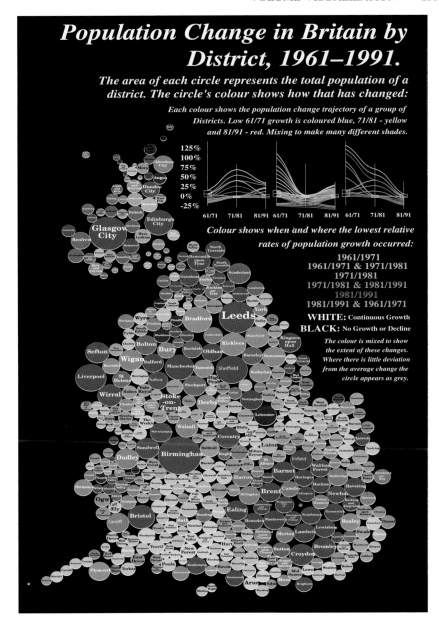

Figure 9.2   This 1981 population cartogram has the 459 districts coloured according to which population trajectory each most closely matched over a 30 year period. The major conurbations have lost population continuously, while the Home Counties ring has mostly always gained people. However, there is much more in the detail. Unfortunately this reproduction cannot match the original distinct set of many more colours.

The Space/Time Trend of
Unemployment in Britain, 1978–1990.

1989/90

Time

(on 1981 County
population
cartograms,
scale indicating
deviation from
space/time
independence.)

1978/79

...Good Times
and Bad...

○ 50% under

◉ 10% under

● 0%

● 10% over

● 50% over

Tyne & Wear

Space

London

Figure 9.3   The 64 counties are each drawn twelve times, shaded by the deviation from the expected rate of unemployment, given the time and the place, on the equal population cartogram using an isometric perspective. Here the pattern for London can be made out, but many other places are obscured. It turns out that crude perspective pictures are an ineffective way of showing a changing situation. Note: the two images shown here are were captured as dumps from a very low resolution computer monitor (illustrating the resolution at which all the images in this book were originally viewed at).

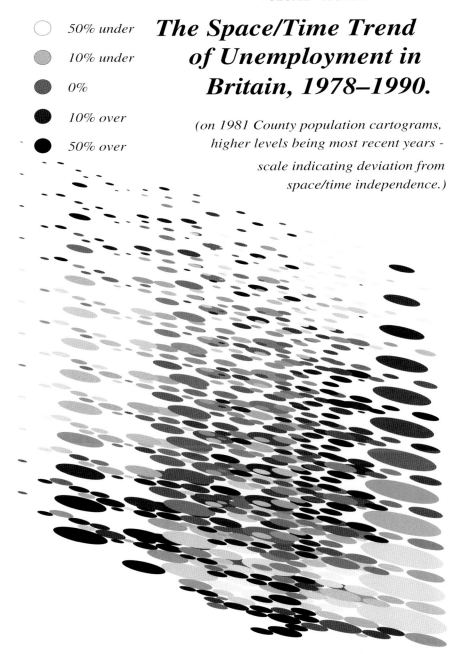

○ *50% under*

◔ *10% under*

● *0%*

● *10% over*

● *50% over*

The Space/Time Trend of Unemployment in Britain, 1978–1990.

(on 1981 County population cartograms, higher levels being most recent years - scale indicating deviation from space/time independence.)

Figure 9.4   This picture shows exactly the same data as the previous one. London can be easily identified on the right as it gets closer each year. This was a particularly disastrous attempt to show the three-dimensional structure of space–time employment change. Frequently you can only find out if a method will produce a satisfactory picture by actually creating the image using real data.

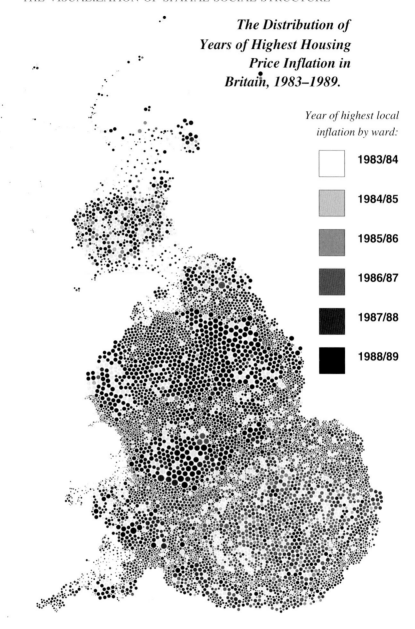

The Distribution of
Years of Highest Housing
Price Inflation in
Britain, 1983–1989.

Year of highest local
inflation by ward:

1983/84

1984/85

1985/86

1986/87

1987/88

1988/89

Figure 9.5   The palest ring can be seen around London where, in 1983/1984, suburban housing prices were increasing fastest. The dark expanse of the northern half of the country shows how prices increased there (but less so) in later years. Scotland appears a relatively independent market. The speckled nature of the picture shows how the situation was not at all uniform; there was great local variation.

and partly because unemployment is known to vary seasonally. The images show dramatic changes in the social structure of Britain.

Initially, in the late 1970s there was only high unemployment in an expanded Celtic fringe, but gradually the picture changed until, by 1990, unemployment was highest in the north and in inner cities, leaving a ring of almost full employment around Outer London. Between those dates, at the height of the early 1980s recession, places like Liverpool were seen to do relatively well as their positions improved in relation to other areas, though it became worse in absolute terms.

Work such as this requires very large matrices of data, over a million rates for the ward time series (which, as a consequence, is not visualized here). It would perhaps be better if the fate of individuals were better known, rather than us knowing these giant matrices of aggregate counts of people's fortunes, but while that is all we have, we must use it. A second problem concerns the use of deviation from the expected, to highlight change. This makes the images seen especially dependent on the first and last years chosen to bound the study period.

In future visualizations it might be better practice to show the changes between individual years, which would be insensitive as to which time span was chosen. Finally, the twelve slices only begin to capture the inside-out goblet shaped structure of spacetime regional inequality in Britain. To see it as it is, we need software that can in a sophisticated way render a projection of a great many observations in space – or a lot of imagination.

9.3   Spacetime continuum

If space and time are considered jointly as a three-dimensional block of space-time with co-ordinates of time, latitude, and longitude, and if incidence (occurrences related to the population at risk) is represented within the volume of the block, it follows that there must be some unevenness.

(Knox, 1964. pp. 20–21)

In some cases, usually when the information is sparse, we do have a near complete record of individual cases. For instance, a complete list of firms opening and closing, including the number of their employees could be created to try and understand unemployment change. The situation where such information is most plentiful is with the medical records of rare diseases. The incidences of when these are detected are available to the level of the address of the sufferer and the day of diagnosis. Such records are relatively few, allowing more sophisticated techniques to be attempted than could be used with the voluminous unemployment data. In Figure 9.6 each case is added to the map as a bell shaped kernel, or cone,[6] the influence of these decaying with distance (Box 9.1) from their incidence in both space and time.

---

[6] The idea of representing incidents by a cone has been previously suggested: 'It may be useful to think of total population influence in yet another way. Each person's influence may be represented

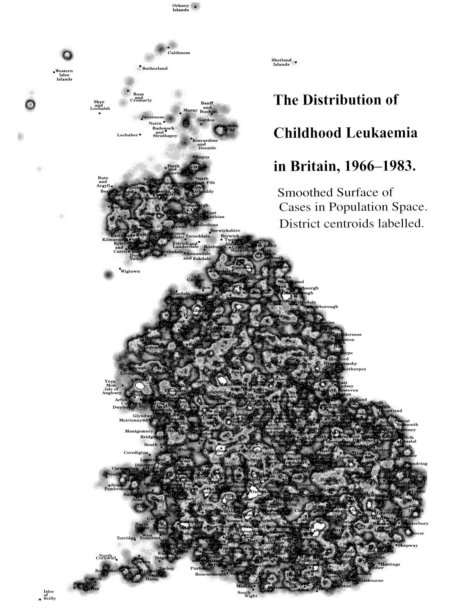

The Distribution of

Childhood Leukaemia

in Britain, 1966–1983.

Smoothed Surface of
Cases in Population Space.
District centroids labelled.

Figure 9.6   The distribution of 9411 incidents of childhood leukaemia (from the National Cancer Registry) are shown as a surface with contours of colour, smoothed by the use of kernels over the equal enumeration district population cartogram. This is an extremely even distribution, with some inevitable random variation. There are no definite clusters shown, but recognizing a random pattern is quite difficult. Note: parts of original image were produced as a bit-map of pixels of colour not as a vector graphics file of lines, curves and areas.

Box 9.1   Three-dimensional smoothing

The one- and two-dimensional binomial smoothing discussed earlier can
easily be extended to work in three
dimensions. The three matrices shown
here give the smoothing factors
around a single *voxel* in three dimensions. After a few passes, application
of this filter approximates the trivariate normal distribution.

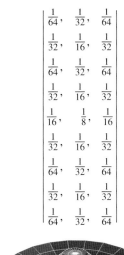

Below is shown a perspective
view of the bivariate normal kernel
which could be used to smooth a point
distribution. A trivariate function was
used to smooth the cancer cases in
this book. The width and sharpness
of the kernel is more important than
the actual shape of the function, which
is chosen for its effect on the final
image. These parameters correspond
to the number of passes of binomial
smoothing undertaken. Often it is useful to explore the effects of a range
of parameters (if possible, while the
image changes before your eyes).

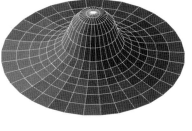

Childhood leukaemia cases in the North of England over the twenty years
prior to 1986 numbered several hundred, with a high level of accuracy involved
in their recording. We can think of these cases as points, sitting geometrically in
a block of two decades of Northern English spacetime (Figure 9.7). If we were
to render these cases as simple points, then, because of their sparsity, we may
well not pick up any slight increase or decrease in the density of cases or some
more subtle spatial and temporal arrangement. To aid visual interpretations here
the points are represented by spheres (Figure 9.8). This process can also create
a truly three-dimensional surface; a single value existing at every point in time
and space is related to the general prevalence of the disease about that time and
around that place.

---

by a pile of sand, with the height of the pile at the place the person occupies and decreasing away
from him. Suppose there is a similar sandpile around the place of residence of every individual.
Now let all this sand be superimposed. At any point the total height of the sand will be the sum of
the heights of all the individual sandpiles. The total height is a measure of total population influence
at that point, and a contour map or a physical model may be made of the entire surface' (Warntz,
1975, p. 77).

To see this space we could again resort to time-slices, in this case taken from an animation showing the development with one frame for every month in the period. However, with only a few hundred cases, more elaborate software can be employed where the spheres are actually placed in an abstract three-dimensional space and a camera is flown around it, recording the views of specific interest (Figures 9.9 and 9.10). Again these are shown here as individual frames, but no longer time-slices, as they were taken at an angle through the block and show an image looking into it, obscured only when cases are eclipsed by one another.[7] As can be seen in the illustrations, the cases are very evenly spread in population space (Figures 9.11 and 9.12).

The actual space in which we place the cases is a very important consideration. A simple Euclidean space has only been used here to show how that image differs from a more useful one obtained when a more appropriate two-dimensional population cartogram is used. Determining the relationship between time and space is not simple. Here it was somewhat arbitrarily chosen to make one year equal to twenty-five kilometres, or the square root of three hundred thousand people.

The idea of a physical distance being proportional to the root of the number of people makes sense as soon as you imagine area, distance squared, being proportion to the number of people, as on a population cartogram. Next you have to think of every person's life having equal volume, not their presence having equal area, and time is included where the relationship between people, time and space has then to be formulised.

The distribution of the childhood population at risk from leukaemia hardly altered over the 1966–1983 study period (Figure 9.13). A more thorough study would have to consider carefully the construction of a volume cartogram, in which every life was equal and placed as close as possible to those with whom it shared life, as well as being near their immediate ancestors and offspring (if any). The relationship between time and space would depend on how far and how frequently people tended to move in their lives.

9.4   Three-dimensional graphs

We need to be able to tell which three-dimensional subspace of the Euclidean data space we are looking at. We also need to see how the point cloud is oriented in that space. To satisfy these needs we draw, in a corner of the screen, an object called the coordinate axes.

(McDonald, 1988, p. 185)

---

[7] Before we begin looking at leukaemia, we can assume, from past research, that we may not find any discernible patterns: 'The limited amount of geographic variation for certain cancers may also provide insights into aetiology. Leukaemia rates were nearly constant across the country, similar to the minor international differences that have been reported. This suggests that the role of environmental exposures may be less important or conspicuous than for other cancers' (Melvyn Howe in Blot and Fraumeni, 1982, p. 190).

Six Views of the Childhood Leukaemia Spacetime Distribution, Northern England 1966–1986.

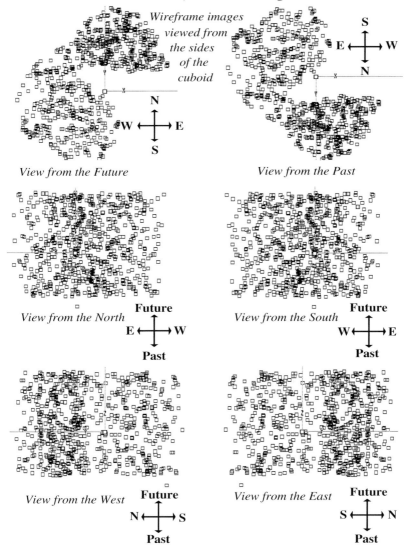

Figure 9.7   Those cases occurring in Northern England alone are shown as wireframe cubes in views through a block of space and time of equal population density. The density of cases is very even in time as well as space. It seems slightly denser in the Newcastle area when looking forward or back in time. However, this impression is not sustained when the block is viewed from the four other faces at right angles.

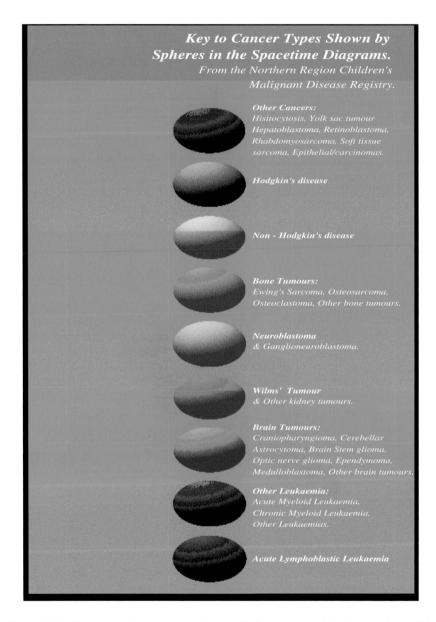

Figure 9.8   Ray-traced spheres are shown, lit from the top left. An artefact of the printing-to-file process (in 1990) occasionally causes darker strips to be drawn across the page. The nine groups of cancers are the amalgamations of the lists shown here. It is not easy to discern more than this number of distinct colours along with the effects of shadows determined by the angle of the light source. Note that these six images were original captured as screen dumps and so the resolutions are relatively low.

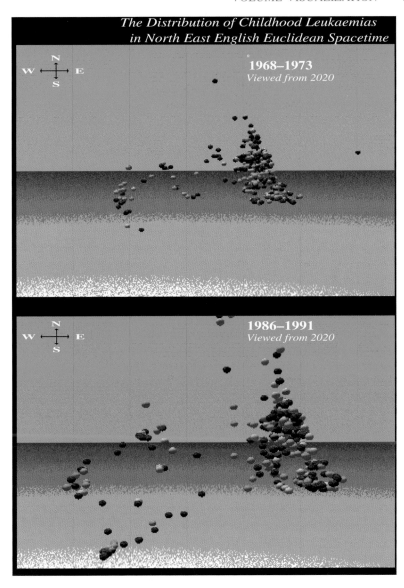

Figure 9.9 *Two slices of time, each of several hundred incidents of cancer, mainly leukaemia, are shown, viewed from a fixed point in the future on the equal area map of North East England, with time as a third dimension. The shape of Tyneside can be made out from the cases. All cases have an equal volume despite the perspective view, which does work by drawing nearer spheres as slightly larger. Note: these two images were originally ray-traced onto a relatively low resolution screen.*

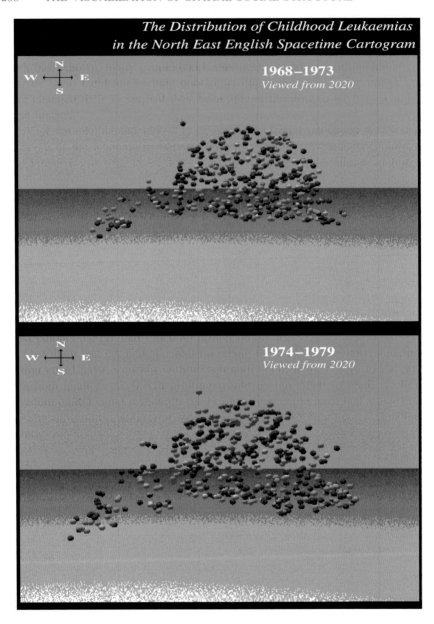

Figure 9.10   The same two slices in time of the same data as the previous figure are shown, but here using an equal population cartogram base. The first time period reveals a very even spread. There may be a ring of childhood cancer cases around Middlesbrough in the late 1970s, but it is difficult to gauge how these incidents fell in time from this angle. The image has to be rotated. Note: parts of this image were created from low resolution computer screen snapshots.

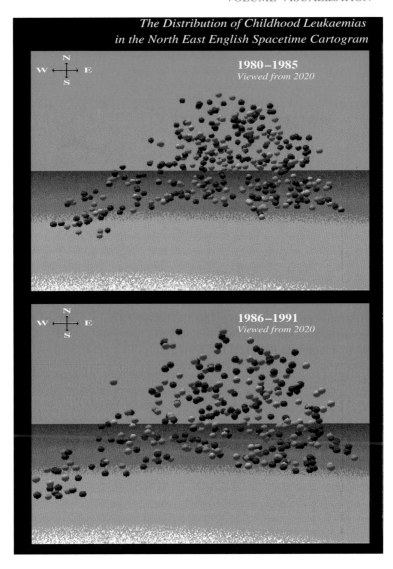

Figure 9.11   A later period is shown here, with similar numbers of incidents of childhood cancer, again on a population cartogram base. The image appears larger than before because the cases are nearer in time, although individually cases are all the same size. Coming closer to the present there are again no clear patterns, but, again, from this angle, a possible grouping of cases in Teeside is seen. Note that the banding of grey-shades at the foot of both these figures is due to the original low screen resolution and limited pallet for colours that the software could use.

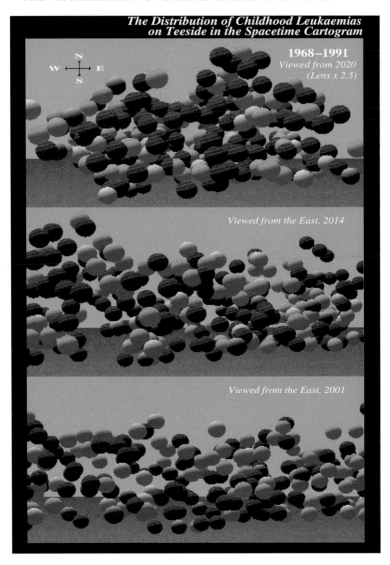

Figure 9.12   Three slices of all the cases of childhood cancer in Teeside are shown as the point of view rotates back in time and towards the east. Rotating the camera around just the cases in one area shows that whenever a cluster is imagined to exist (and many thought they did exist at that time) it almost always disappears as the angle, and hence the view, is altered. The light source was in the west. Note that the lines that appear across many of the spheres partly reflect the limitations of early visualization software but should not detract from the purpose of the graphics.

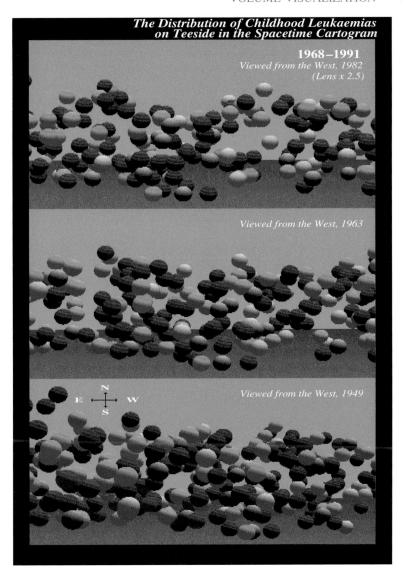

Figure 9.13   Again, three slices of all the cases in Teeside are shown as the point of view rotates back in time and towards the west. As before, what appears to be a pattern in two dimensions often breaks up given a different three-dimensional view. This distribution is so even, and the number of cases so low, that any conclusion drawn about a possible pattern would be extremely tenuous. Note that the relatively low resolution of these three images is a product of the original circumstances under which they were created.

Things other than space and time can be projected to occupy three dimensions. A three-dimensional graph is created by merely raising a third axis at right angles to the conventional two, and plotting points inside that space. A '3D' graph is not therefore what the term is commonly used to describe, a one-dimensional bar chart, with each bar drawn as a pillar.

The three-dimensional graph used here is the natural extension of the equilateral electoral triangle (see Chapter 7) developed into the logical three-dimensional analogue of a tetrahedron, which attempts to show how the vote is shared between as many as four parties in a large number of areas (Box 9.2). We need to be able to do this if we are to include Scotland in our analysis of electoral composition in Britain[8] (Figure 9.14). In Scotland, in recent years, the Scottish Nationalist Party

Box 9.2   The electoral tetrahedron

The idea of the equilateral triangle can be extended into three dimensions in a tetrahedron to show the composition of the votes of four parties, among a number of constituencies. Position (x, y, z) on the triangle is calculated from the Conservative (C), Labour (L), Liberal/Alliance (A) and Nationalist (N) proportions of the vote as follows:

$$C + L + A + N = 1$$

$$x = \frac{(C - L)}{2}, \quad y = \frac{(A + N)\sqrt{3}}{2}, \quad z = N\sqrt{\frac{5}{12}}$$

Position in the equilateral tetrahedron formed then gives the share of the votes in any one constituency and a map of them gives the distribution of all constituencies simultaneously. To understand the distribution within the three-dimensional space it must be rotated by the viewer. A two-dimensional net of the space can be opened out to expose some of the pattern on flat paper, but a lot of the dynamism of the graphic is lost. It

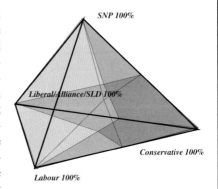

is hard to imagine how this device could be profitably extended to show the composition of the vote among five parties. Three dimensions are hard enough to grasp.

[8] In 1991 it was possible to write 'Scottish election results have recently become a truly four-way affair'; currently (in 2012 as I make the last changes to this text) the SNP are in power in Scotland and promise a referendum on independence. Things were changing two decades ago: 'Two-way contests, which were far and away the most common in 1974, have declined pretty steadily and

(SNP) has consistently come in third or second place (by 2011 it was placed first in Scotland), but the third major British party is still in the reckoning and has quite a different pattern of support.

The abstract creation of an electoral tetrahedron is quite simple. The points, representing the electoral divisions in which the vote is counted, are placed so that their distance from the four apexes is in exact proportion to the share of the four-party vote represented by those apexes. To stick to convention, looking from above, the Conservatives have the right hand apex, Labour the left, the Scottish Liberal Democrat (SLD) party the top (still on the plane) and the Nationalists have the apex in the centre (now above the other three).

The problems are familiar when we attempt to project the three-dimensional structure of voting on to a two-dimensional plane for drawing. Here the method involves taking slides from an animation and showing those slices from particular angles. However, another method has also been developed in this case. That is to unfold a net of the tetrahedron as four equilateral triangles (Figure 9.15). A point is drawn on the side of the tetrahedron it was originally closest to.

As the centre of the political battleground tends to be relatively empty (due perhaps to tactical voting), the unfolded tetrahedron technique does not create results that are too ambiguous.[9] In fact, each triangle in the net contains only those divisions where a particular one of the four parties came last. The net can actually be further subdivided into areas in which the exact order of the parties is known. Such an arrangement makes interpretation of a complex three-dimensional situation considerably simpler (Figure 9.16), although in explicitly using two dimensions something has to be lost – in this case the fortunes of the party coming fourth in each constituency.[10] The four triangles show the positions of the parties coming first, second and third. Along the six edges are shown the relatives positions of the parties that come first and second. It is clear that the most common result is Labour first, SNP second.

---

significantly. In particular, straight fights between Labour and the Conservatives, which were again the most common, are now relatively rare. The increase in Conservative v Labour v SNP contests is a direct function of the larger number of SNP candidates. This also explains why, despite the fall in the number of SLD candidates, the proportion of four-way contests reached a high point of twenty-three per cent of contests in 1988' (Bochel and Denver, 1988, p. v).

[9] Movement to the apexes of the electoral tetrahedron would indicate that the following had occurred: 'This apparent consolidation of strength in the parties' own territories is an interesting phenomenon; it is unclear on the available evidence whether incumbency of itself gives an advantage or whether parties successfully targeted their campaign effort to exploit existing support' (Bochel and Denver, 1990, p. vi). In hindsight what was really happening was voting polarisation following social polarisation.

[10] How reliable are our visual and mental abilities when dealing with this complexity?: 'The use of such a system poses interesting theoretical questions: Is exploring data by looking at projections "safe" – if you look at enough different projections of structureless data, will you find structure by chance? If it is safe, is it "effective"? – in what sense can the information in a d-dimensional point cloud be extracted from a few of its 3-dimensional projections? The method, properly applied, appears to be both safe and effective, even allowing for the fact that we do not know the statistical properties of the eye as pattern detector' (Donoho et al., 1988, p. 119).

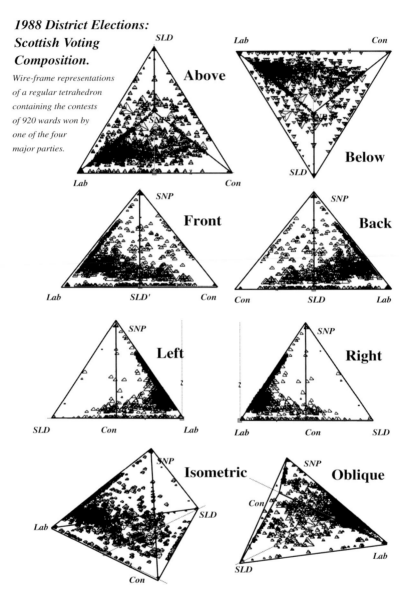

1988 District Elections: Scottish Voting Composition.

Wire-frame representations of a regular tetrahedron containing the contests of 920 wards won by one of the four major parties.

Figure 9.14 Once there are more than three main parties in an election, you need a tetrahedron, and on paper a set of views, to show fully the voting split. Otherwise all but two of the parties have to be 'Others'. Here the centre of the tetrahedron is generally avoided and wards tend to cluster along the SNP/Labour and Labour/Conservative axes, with low Scottish Liberal voting at this time. Note that the eight views were all originally screen-shots and so the resolution of each is not especially high.

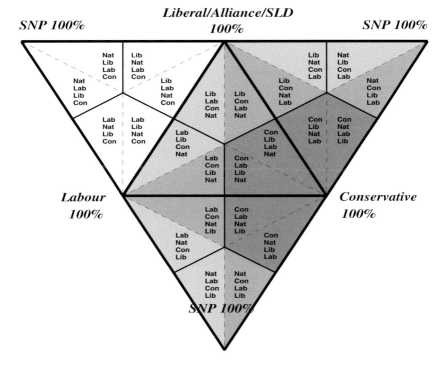

Schematic Representation of Four Party Voting Compositions.

The diagram below is a projection of the tetrahedron which contains all possible four party arrangements. The standing of the Conservative party is shown as an example in shades of grey. Position on the diagram shows party order as well as proportions of the vote.

Figure 9.15   An equilateral tetrahedron is unfolded to illustrate one possible way of showing the local voting distribution in Scotland. Each triangle shows one of the 24 possible orders of rank for the four parties. Here they are shaded to show where the Conservatives came first, second, third or fourth. The central triangle is where the SNP were placed fourth (labelled as 'Nat' for Nationalists here).

1988 District Elections: Scottish Voting Composition

The triangles show the projection of a regular tetrahedron encompassing electoral ward competitions involving as many as four separate candidates. Every ward won by one of the major parties is shown as a circle on the diagram, its area in proportion to the total vote. The position of the circle indicates the composition of votes in that ward. Circles are shown on the side of the tetrahedron they lie closest to. Wards falling on the edges of the tetrahedron are projected as histograms of two party support on the sides of the triangles. Distance from each apex measures the support for a party from total to none.

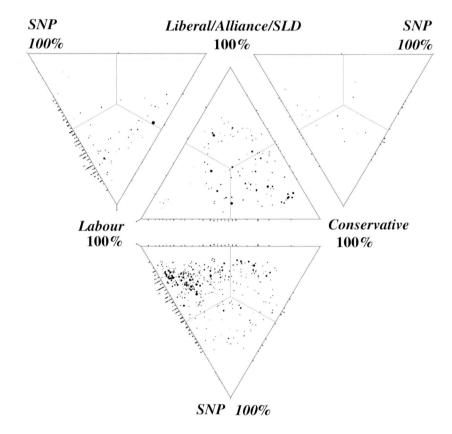

Figure 9.16   Using the previous layout, every constituency result is shown here on a three-party triangle, which ignores any party that came fourth or worse. Two-party contests occur (twice each) on the sides of the triangles and each uncontested seat (all then won by Labour) are at the three corners. The most common three-party result was Labour first, SNP second, Conservatives third and Liberals fourth.

What do we do, though, if we wish to see how the four-way situation changed over time? A slight change in the number of votes could send a division spinning across the net, which in reality would hardly move in the tetrahedral space. This would be unfortunate, unless only changes of party position were of interest.

What if there were a fifth party also of some importance, for example if the green vote rose up in the future? Could we show a two-dimensional net of the three-dimensional shadow of a four-dimensional hyper-tetrahedron? Or would it be better to observe the rotating three-dimensional net of the four-dimensional point cloud (Figures 9.17 and 9.18)? These situations are avoided for the while, but remain to be addressed.

9.5   Flows through time

Rather than trying to simply display the data the idea is to extract certain topological information and to display this ... a jillion little arrows displayed in a cube would not reveal much about a three dimensional flow.

(Nielson, Shriver and Rosenblum, 1990, p. 261)

If the possibilities at the end of the last section appeared a little daunting, consider, for a moment, the problem of showing how a pattern of flows has changed over a number of years – not a single change but a complete succession, ideally going back some way in time.[11] Just such a truly three-dimensional matrix has been constructed from the National Health Service central register, for flows (in both directions) between every pair of ninety-seven mainland family practitioner committee areas for each of the last fourteen years.[12] Even this low level of spatial and temporal resolution produces over one hundred thousand separate counts of migration streams. How can we begin to see what is happening to the flows of people that create the spacetime pockets of existence (we call places) in Britain?

---

[11] When you go back a little further in time you can easily be surprised. 'Particularly impressive was the way in which the South East's population began to decline in the late 1960s, following its rapid growth in the 1950s and the early 1960s' (Champion, 1989, p. 122).

[12] Migration patterns have been fairly consistent over time, but do fluctuate; without images we tend not to know much more than all these numbers: 'In 1989, the total number of moves between FPC areas within England and Wales, at 1.76 million, was 6 per cent less than the 1.88 million in 1988 .... There was little variation in the total number of moves during the years 1979 to 1985, which ranged from 1.50 million (in 1981) to 1.60 million (in 1985). However, in 1986 the number of moves increased to 1.83 million (a 14 per cent increase over 1985), with further increases to 1.87 million in 1987 and 1.88 million in 1988 .... During this period, expansion of financial services, resulting in easier access to mortgages, and relatively low interest rates may have contributed to the increased number of moves. Similarly, the fall in the number of moves in 1989 could have been partly due to the rise in interest rates. The total number of moves in 1989 was still 13 per cent above the average for the seven years before 1986' (Bulusu, 1990, p. 33).

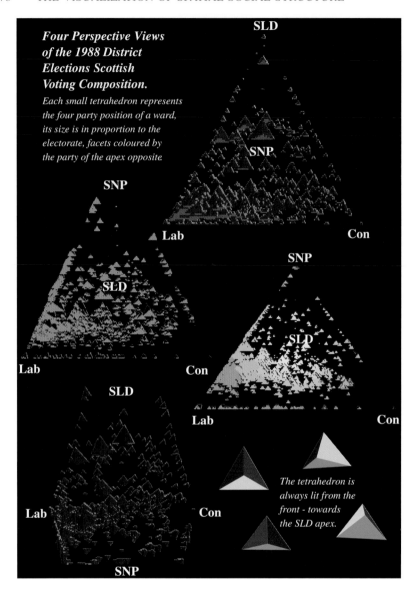

Four Perspective Views of the 1988 District Elections Scottish Voting Composition.

Each small tetrahedron represents the four party position of a ward, its size is in proportion to the electorate, facets coloured by the party of the apex opposite.

The tetrahedron is always lit from the front - towards the SLD apex.

Figure 9.17   Colour can be a useful cue in seeing inside three-dimensional structures. Here the tetrahedrons representing the wards are shaded by the colour of the party on their facet opposite the party's apex. The view is lit towards the SLD (Liberal) apex. The perspective projection produces greater magnification of near wards. It is best to spin these types of image on a computer (or other) screen. Note that parts of these images appear to be at especially low resolution because the algorithms used to colour each triangular facet of the tetrahedra do not map colour well onto large square pixels.

**A Ray-traced Image of the 1988
Scottish District Elections Voting
Composition.**

SNP

SLD

LAB

*The four party tetrahedron
viewed from just above the
Conservative apex.*

CON

*Figure 9.18   This image shows something of the empty hemisphere in the core
of the four-way distribution. The line of tetrahedrons in the foreground represents
seats contested only by the SNP and Conservative parties. The lines of other two-
party distributions can also be discerned. Even had the image been at a higher
resolution, little more would have been obvious without rotating it in space. Note
that in the foreground the tetrahedra appear as very jagged due to a combination
of low original screen resolution, a small colour pallet and the angle at which
light hits each surface.*

The basic two-dimensional flow maps showed numerous overlapping arrows. When the change between two years was sought, even depicting the single differences could require two separate images. Depiction of spacetime flows of people would have to be constructed in three dimensions. Theoretically there would be a plane to represent every year, which would contain the changing population cartogram. Places would be linked by tubes, the width of which, say, was in proportion to the number of migrants. To prevent the image becoming too tangled in practice, a measure of significant change would have to be found, similar to that used in the two-dimensional case. Otherwise almost ten thousand tubes would have to connect every pair of time planes. At least the origin and destinations of migration would be obvious, even if the paths between them were, to say the least, a little confused (Box 9.3).

The structure just described has not been created here; maybe it would just produce a perplexing mess. To understand such a structure, even after generalisation, requires the development of new techniques to look into three dimensions. A maze of tubes crisscrossing in spacetime will not reveal its structure through the illumination of its outside surfaces. Cross-sections through the connections would be confusing, and it is also not possible to simplify such a complex organisation to a plane and retain its essential form.

Box 9.3   Three-dimensional structure

A complex three-dimensional structure is sure to appear extremely confusing when forced to fit on flat paper. These two graphics show an experiment to project the spacetime distribution of unemployment and the use of tubes to show migration flows across space and time.

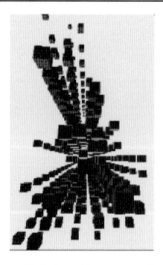

Eventually being able to rotate these images is not enough. We need to be able to get inside them to explore and discover what the structure to the patterns may be. With 1980s home computer graphics, we could only paint pictures of the outside and were unable to see within.

9.6   Volume rendering

It is certainly feasible, and may prove useful, to offer a 'biod's eye'
view of the dataset as viewed by one of the biods, using stereoscopic
viewing and other 'virtual reality' techniques as they develop.
(Kerlick, 1990, p. 127; 'biod' is 'bird' and 'icon' combined)

So far this chapter has concentrated on rendering volume visualization, not
looking into the volume. That is because what most of the facilities available
in quantitative social science can at best achieve is only to show surface views
of three-dimensional structures.[13] That is why only this chapter of this book has
been devoted to the subject which, in the late 1980s and early 1990s, was the
concern, above all others, of most interest in computer science visualization.
What was then new software was being written to look inside the surfaces, to
create images on the screen that we could not see in a picture on paper. Volume
rendering defines what can be done with this kind of software, which can only
be described in these pages; it cannot be shown.

The key theme is translucence. Surfaces can be peeled off a volume, like
two-dimensional contours, but really to see the structure you must be able to
see all the contours at once. To do that objects must emit and transmit light, so
that they can be seen through, but also not be transparent, in order to still be
seen themselves. There is obviously a limit to how many layers can be pierced;
each obscures a little more than the last. The combination of translucence with
perspective, animation and lighting allows us almost to see inside the volume as
we move around and through it.

Imagine the economic spacetime of Britain with the unemployed areas shown
like dark storm clouds through which it is possible to see better times ahead
(in time) or to the side (in space). The whole structure would be held in the
holographic image where no pocket of prosperity or despair could remain hidden.
What would the spacetime continuum of childhood leukaemia incidence look like
seen through that translucent space? When these words were originally written,
translucency was cutting edge. Now, for much basic everyday software, it is
the common format in the most widely used microcomputer windows operating
system used by billions worldwide.

In translucent space no case could ever quite eclipse another.[14] More importantly, when two or more cases fall in almost exactly the same time and place

---

[13] The old approach was to show three-dimensional structures through two-dimensional surfaces:
'The second, newer approach to volume visualization is called direct volume rendering, volume
imaging, direct voxel rendering, or just volume rendering. This approach maintains an explicit connection between the volume data set and the volume visualization. The algorithms use no intermediate
geometric representation. The resulting voxel clouds, perhaps more visually ambiguous, permit users
to explore directly the contents of their data. The scientist can slice-and-dice the visualization to
explore arbitrary cross-sections of the original volume data set. Viewing is not limited to surfaces,
although surfaces are sometimes portrayed' (Herr, 1990, pp. 201–202).

[14] Great claims were made for the future of interactive computer graphics in the 1980s, especially
for volume visualization where the volume changed shape (called here 4D), but also for looking at

they will appear much darker than is usual, rather than as a single occurrence. The three-dimensional electoral graphs would appear more like a cloud of dust particles or a galaxy. The true density and sparsity of spatial divisions would be apparent where, before, they had quickly obscured each other as a dark mass. Lastly, there are the flows through time, depicting those so that it might be possible to trace the path of each migration stream through the myriad structure of pipes and columns.

Translucence is not true three-dimensional imaging. To argue that is rather like telling a two-dimensional being that can only see a one-dimensional strip that, if objects were made translucent, it would see two-dimensional structures. It would not. It would merely see what was previously completely hidden from its view, and the viewer may, through rotating the angle and position from which it viewed the two-dimensional space, come to guess some of its structure. However, it would never have the full luxury of being able to see simultaneously all that it contains and how it is arranged from above, because it is part of that two-dimensional space. Similarly, in the real world, we will never have that visual ability in three dimensions.

9.7   Interactive visualization

Just as 'visualization' has been invented to describe the process of providing more immediate access to very large amounts of data, 'interactive visualization' will be 'invented' to describe the process of providing more immediate access to the particular features that are of interest to the analyst at particular points in both the spatial and time domains of a given field.

(Dickinson, 1989, p. 10)

To bring the discussion up to date, in the late 1980s it was being claimed increasingly that there were two types of visualization – the mundane variety, which would include this book, and the interactive kind, the most extreme example of which is found in the artificial realities of computer graphics.

Interactive visualization, like interactive graphics, allows the viewers to change instantly, but smoothly, the direction and position from which they are viewing, what they are seeing and how it is depicted, lit, animated and so on. What you see moves, and so can you.

Freedom to interact allows any aspect of a structure to be examined at will. It is almost as if you could pick it up and turn it around in your hands.

'survey results', by which these authors mean aspects of social science. As yet almost no such '4D' social science visualization has been achieved: 'Interactive computer graphics is the most important means of producing pictures since the invention of photography and television; it has the added advantage that, with the computer, we can make pictures not only of concrete, "real world" objects but also of abstract, synthetic objects, such as mathematical surfaces in 4D ... and of data that have no inherent geometry, such as survey results' (Foley *et al.*, 1990, p. 3).

In some systems you see the object stereoscopically through two images in a pair of goggles – better still, etched on contact lenses,[15] Your wishes are executed through the movements of your head, hands and even entire body. For the majority of researchers, however, let alone students studying, interactive visualization will not arrive for several years yet.

The basic questions of what it is we wish to see and how that should be drawn remain as important as ever. Interactive computer graphics will allow you to pick up the earth in your hand and view it just as if it were a real globe[16] – but we can already do that in the classroom with the plastic model. What is exciting about visualization is the facility it offers for us to transform what we wish to observe to a form most amenable to our understanding and then change that, if it does not suit us, at a whim.

Interactive visualization will reach the microcomputer screen by first offering the user the ability to link several displays of the same data to gain greater insight – say a rotating tetrahedron of the Scottish voting composition in one window coupled with an animated cartogram in another, showing how the distribution of divisions changed geographically over time. An area of Glasgow in the 1970s could be selected and the points representing those divisions would light up in the tetrahedron simultaneously. As you moved a pointer over the changing cartogram of Scottish divisions, other points would become lit and you could trace patterns between geographical, historical and political spaces.

Artificial reality allows us to go one step further: to be actually inside the tetrahedron; to look in a spacetime cartogram down at the 1986 regional election results and up at the 1990 contest; to see the dark clouds of unemployment rising above Glasgow to meet the fine detail of the 1991 census in the distant sky. Even further above in the far distance are the translucent mosaics formed from what we know given the 2001 and 2011 censuses. Below us would lie the remains of decades of industrial structure and behind us the same for England and Wales. Here 'up' is the future and 'down' is the past.[17] All this will require considerable imaginative leaps, more research and a great deal of development, but if we cannot specify our aims at this stage, how can we plan for the future?

---

[15] Rather than wear goggles containing visual displays: 'An alternative design would be to fabricate a display on a contact lens and a sensor would detect eye movements as well as head and body movements. This display must then generate the image that the eye would see. Since it would only need to illustrate the small area that the fovea would see, the resolution of the image could be very modest' (Krueger, 1983, p. 100).

[16] This is now possible with 'Google Earth', which, as predicted, does not allow the surface of the earth to be transformed (yet), for instance, to be a world population cartogram.

[17] Different people have different perceptions of what is 'up' and what is 'down' in these virtual realties. For some the past may be above and they are descending into the future. Place people in the shadow (or net) of four-dimensional space and, just as in the tesseract, it is possible for some to be walking on what looks to others to be the ceiling. There is a tesseract spinning at the bottom of my website: http://www.dannydorling.org/as a reminder that we have yet to visualize using such things in social science. Click on it for more details. The artist Escher used ideas such as this in his woodcuts. For the ideas behind Escher's images you can start at: http://en.wikipedia.org/wiki/M._C._Escher.

Whether the technical innovations suggested here will result in useful still pictures for printed work (such as this) also remains to be seen. What we do now know is that just because something may appear to be a good idea that alone is not enough for it to be taken up and implemented. In the social sciences most of us stopped looking at the routine data, just as that data began to become interesting, showing patterns that were no longer routine, but of a newly dividing country. This occurred just as we could have begun to look at it in more interesting ways and then shown others what was occurring.

10

Conclusion: Another geography

And let us be aware of the natural tendency to design computer-graphic programs to imitate traditional cartographic conventions in the mistaken belief that because they are traditional they are, ipso facto, legible.

(Bickmore, 1975, p. 350)

New techniques have been presented in this book exploiting the visual processing ability of the human brain to provide an opportunity for interpretations that may better capture the nature of society.[1] This is unlike black box techniques, in which the last thing you can see is what is really going on. In simple statistical analysis, for instance, sharp lines are often drawn between what is significant and what is not, giving little insight into the true nature of a picture of numerous interactions.

By placing people in their spatial context, we often find unforeseen patterns of great interest (Figure 10.1). Visualization allows for a more open-minded style of analysis. It may seem as if we are doing little more than observing the world, but that world is usually hidden from us in vast tables of facts and figures, or in

---

[1] 'Perhaps one day high-resolution computer visualizations, which combine slightly abstracted representations along with dynamic and animated flatland, will lighten the laborious complexity of encodings – and yet still capture some worthwhile part of the subtlety of the human itinerary' (Tufte, 1990, p. 119).

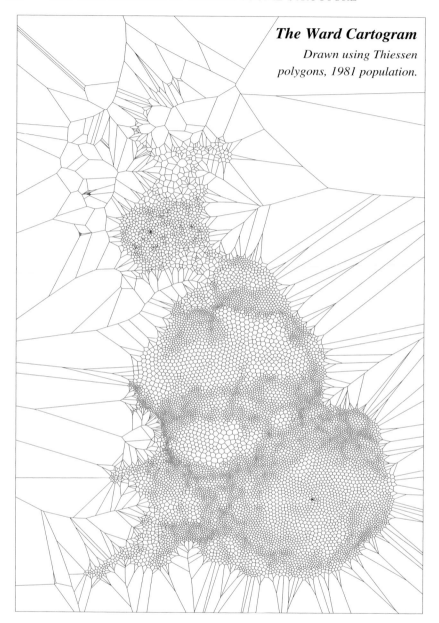

Figure 10.1   Thiessen polygons constructed around centres of 10 444 census wards from the 1981 census equal population cartogram. Space is subdivided by lines bisecting unseen lines joining neighbouring invisible centroids. The resulting image is quite elegant. Time can also be divided in a similar manner. There is a problem with the temporal as with the coastal boundary, which is well illustrated here.

database driven geographic information systems.[2] The best instrument we possess through which to acquire and assemble such information is our visual processing ability through which our imaginations can work to construct knowledge.[3]

When computers were first introduced to social scientists a contradictory position was often held. The computer would soon be able to comprehend the world, to see it. The machines could then tell us what was going on. Only later did it ironically become obvious just how powerful our own human visual perception is and how difficult it is to get a computer to mimic vision.

Writers in the past have thought of, and asked for, ideas that have been brought to a rudimentary level of fruition here – a new look at cartography, more productive employment of graphics, harnessing the machine's power and the mind's intelligence.[4] It is hoped that this book goes a little way in showing how to achieve many of those wishes (Bunge, 1975; Angel and Hyman, 1972; Pred, 1984). The writings and drawings of a great many authors were consulted to try to ensure that no substantial contributions made prior to 1991 were overlooked. Many will have been overlooked that were made since, but as the popularity of using data in social science fell so quickly after 1990, so too the number of overlooked contributions will have been lower than would otherwise have been the case.

The production of new forms of, and uses for, area cartograms is one novel contribution of this book and most of those included here have not been put in print before now. Others look a little like a very rudimentary form of the new types of cartograms that are now more widely available. Those used here were developed with function foremost in mind (Figure 10.2). These took the longest time to create and the algorithm was not easy to develop. Its implementation was actually achieved using computer graphic techniques (Figures 10.3 and 10.4).

The dissertation from which this book is derived was one of the first to work visually with social information of such magnitude and detail, overcoming many of the problems often said to make the handling of so much information

---

[2] Early on it was realised that geographic information systems would only answer simple questions, not help us think about more complex problems, as maps can: 'A computer bank would probably be more geared to answering some specific question *ad hoc* and perhaps less to provoking thought about what questions should be answered' (Bickmore, 1975, p. 344).

[3] 'For the human imagination, always too limited, always curbed by socio-cultural contexts, map collections present possibilities as vast as the data bank is large. Visual selection is faster and better than any automatic selection, since it permits from the outset a variety of nuances beyond the capability of any computer. But its costs in terms of time only pays off with "seeing maps." "Reading maps" make the operation impossible' (Bertin, 1981, p. 161).

[4] Using faces to allow people to see patterns in data was especially ironic: 'This approach is an amusing reversal of a common one in artificial intelligence. Instead of using machines to discriminate between human faces by reducing them to numbers, we discriminate between numbers by using the machine to do the brute labor of drawing faces and leaving the intelligence to the humans, who are still more flexible and clever' (Chernoff, 1973, pp. 365–366).

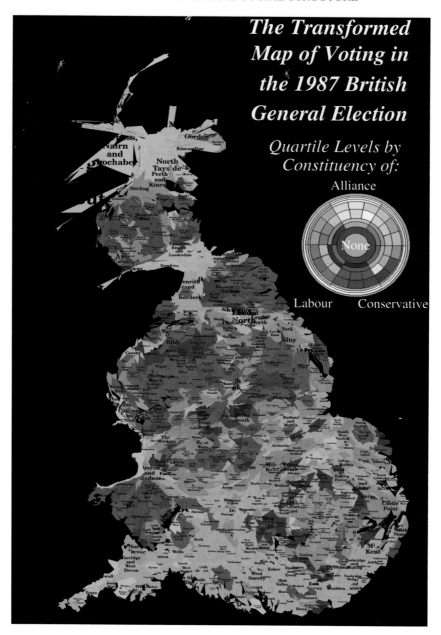

Figure 10.2   One frame taken from the end of an animation showing the 633 mainland parliamentary constituencies, coloured by the mix of voting in 1987, having been transformed from an equal land area to an equal population projection. Here the actual boundaries of the constituencies have been transformed. This produces a more continuous area cartogram. There is better software to achieve this today.

impossible.[5] Much of the practical side of this work was only feasible due to the use of what were then very recent advances in computer hardware and software, but also the arduous collection of large amounts of digital information, which remains a very boring task today (Figures 10.5 and 10.6).

Here the abandonment of many past practices is suggested, accepting that all methods, the ones advocated here included, are tied to the times and places in which they were created, and can never be universally appropriate. You have to have an open mind to be able to accept viewing space and time warped in unfamiliar ways. We need to see more flair in research.

Visualization, it must be stressed, is much more than pretty pictures.[6] It is a methodology for visually modelling aspects of our world to gain a new, useful and different understanding. In hindsight this is not so true of the 1990s, but is truer today, as we are rapidly moving into an age that will manufacture artificial realities from new kinds of liquid crystal screen kaleidoscopes to computer games unimaginable three decades ago, to fictional fractal lands, worlds and galaxies.

The imaginative escape from reality is accelerating. We must be careful that we do not ignore the real world while creating so many artificial universes, careful not to live too much in artificial social spaces, living in Facebook space rather than meeting face-to-face. Be careful not to talk too much by email, rather than in person. In person our eyes allow us to gauge reaction as we both speak and listen; vision has always been a part of talking, so that when the telephone was invented in the 1870s, we had to make a step change in our ability to visualize who we were talking to.

Different goals lay beneath the surface of the 1980s visualization revolution. At a superficial level, there was the aim to extract more money from the academic funding agencies of America.[7] This was carried out with the threat that the Japanese would win the new economic war, and a country already sliding down the world scale must counter-attack. In hindsight we know that the Japanese

---

[5] We are still learning to use some visual interfaces twenty years after they were designed: 'Think of these computer models and the windows provided to the models by the graphics systems as the basic primary representation of information. Not many people have such systems. They are light years away, not because we don't know how to build them, but because we don't know how to use them' (Evans, 1973, p. 7). Today these windows are here – on every desktop – but what are we using them for?

[6] Visualization provides a very different paradigm to graphic design: 'These examples reveal an unconventional design strategy: "To clarify, add detail." This strategy works because humans are well-equipped to deal with masses of data. Massive structures fill our world (we see the tree rather than count the leaves), and the presence of micro information allows viewers to select their own level of detail, picking out the data important to them. This contradicts a commonly held view that data display should be reduced to poster-like simplicity, which imposes the designer's view on the data and limits the usefulness of the graphic. Ultimately, we need complex displays of data because of the complexity of the world being modelled' (Freeman, 1991, pp. 113–114).

[7] The threat of Japanese supremacy is well worn: 'Laurin Herr, an analyst with Pacific Interface, New York, cautions, "In '83 we saw the first wave of Japanese vendors, but they were second tier vendors." Although they did not penetrate the U.S. market then, they are a threat and Japanese devices are already inside our workstations, he says' (Frenkel, 1988, p. 113).

Figure 10.3   Seven views of the progress of a computer animation that created the Northern half of the previous figure. Notice how much the map changes from green and blue domination to a cartogram of almost total red and purple and a little yellow/green: the Liberal and Scottish National Party regions shrink, while the Labour and Conservative/Labour battleground grows in proportion to voters. Note: parts of original image were produced as a bit-map of pixels of colour not as a vector graphics file of lines, curves and areas.

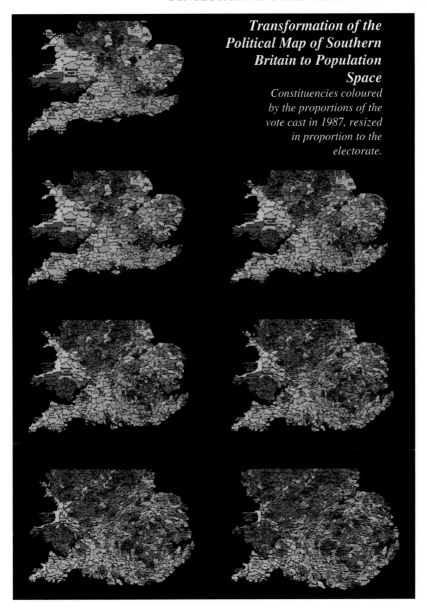

Transformation of the
Political Map of Southern
Britain to Population
Space
Constituencies coloured
by the proportions of the
vote cast in 1987, resized
in proportion to the
electorate.

Figure 10.4   The same process is shown for the southern half of Britain. Here
the distorting impression given by the original map is clear. The true red, blue
and green polarisation becomes apparent. The division of the southern half of
the country into green and blue areas, stretched around red and purple places,
becomes obvious, while a bright slice of yellow still remains on the Celtic fringe.
Note that these images were taken as seven screen shots from images designed to
be seen on one of the very first colour graphics computer monitors made for a
home computer.

Figure 10.5   The 650 parliamentary constituencies shown on the electoral triangle (with Northern Ireland shown separately), shaded here by the level of turnout. It is clear that the highest turnout is to be found in marginal and Conservative party seats, although a few bright spots can be seen in some Labour strongholds. In Northern Ireland turnout was generally low, as it was in the few Alliance strongholds.

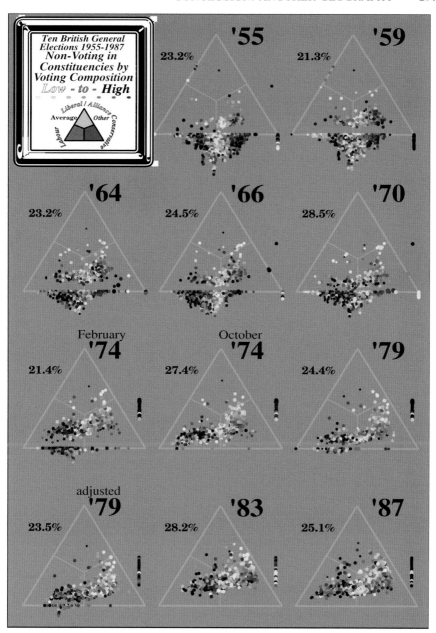

Figure 10.6   In the 1950s and 1960s, most marginal seats were very strongly contested, shown here in pale shades, indicating high turnout. In 1970 the safest Conservative and Labour seats also had a high turnout. During the 1970s the composition is seen to change and the pattern of turnout alters to reflect this swing. The Northern Irish seats move out of the triangle as new parties appear there.

economy hit an output peak on 29 December 1989 and then began the long 'lost decade' that rolled into two decades, although, post 2008 crash, maybe these decades were not so much 'lost' as 'realistic'. It is easily possible to be producing too much, just as it is possible to try to control too much.

There was far more behind the first visualization revolution in academia than new technology and the hunt for economic power. Maps and charts, by containing information, have always been a key to power and social control. Their origins in military conquest are replicated today by efforts in (spy satellite) image processing and (battle ground control) geographic information systems. Just as the clock allowed the timing of people's lives to be controlled, so the map permitted regulation of spatial movement and enclosure of land.[8] Today, these are combined in information systems which, with visual capability, create new possibilities for technocratic control, through determining the accepted image of the world.

Another aspect of visualization is seen though the possibilities it holds to reveal the injustice and inequalities in the world and to show these pictures to more than just the bureaucrats and administrators.[9] Images are becoming the currency of the information age. We are now used to receiving much of our understanding through the flat screen. Despite publishing more, we are currently reading less and viewing more.

To communicate we must compete with others' graphics. How better than through our own graphics? This would be another step change in how humans communicate and show displeasure, agreement and contributory or conflicting ideas. We have always drawn images, from before we could write, but what if we could draw so much faster and clearer than before?

Showing different images of the world by using new ways of depicting the information we have has always caused fear. For instance, there was once a time when the simple traditional Mercator map was viewed as a little too powerful an information transmission device to allow the knowledge it presented to be shared with ordinary mortals, especially with school children. Geographical knowledge has been long recognised for the power it offers. Prior to 1900 in England, it was said that:

---

[8] The history of the use of maps for social control is well documented: 'Maps impinged invisibly on the daily lives of ordinary people. Just as the clock, as a graphic symbol of centralised political authority, brought "time discipline" into the rhythms of industrial workers, so too the lines on maps, dictators of a new agrarian topography, introduced a dimension of "space discipline". In European peasant societies, former commons were now subdivided and allotted, with the help of maps, and in the "wilderness" of former Indian lands in North America, boundary lines on the map were a medium of appropriation which those unlearned in geometrical survey methods found impossible to challenge' (Harley, 1988, p. 285).

[9] Technical change brings little that is fundamentally different: 'Are we returning to a new Dark Ages? Will the GIS specialists become the new priestly class, determining our image of the world just as surely as did the makers of the MAPPAE MUNDI?' (Harley, 1990, p. 15).

Such education as there was did not include the dangerous subject of geography; even in the National and British schools of the period (to say nothing of the workhouse schools) there was such a prejudice against the teaching of geography that in many cases the school master was forbidden to hang any maps on the walls of the schoolroom.

(Redford, 1976, p. 96)

The social conclusions of the research underlying the examples used in this book are that British society had become sharply divided by the late 1980s (Figure 10.7). The divisions were most obvious when viewed on a fine spatial scale, when the cartographic microscope of a detail cartogram could be employed. It was at this point that it became evident how people were socially herded into, or could not escape from, many areas – for many reasons.[10] It was evident back then that there was little reason to think that these divisions would not widen in the future. There appeared to be nothing likely to curtail this 1980s polarisation quickly once it had begun, particularly when the people in more prosperous places held the political power.

Even if in the late 1980s we could not easily have changed what was happening, we could at least have shown it better for what it is.[11] Nothing ever changes while people cannot sense what is happening. People can be hidden in the detail of conventional maps. Many of the origins of our current malaise were not seen in traditional equal-land-area cartographic images of British lives lived in the 1970s and 1980s.

It is perhaps the extent of the problem of current social inequality that often helps prevent its full appreciation and dissuades us from taking action.[12] Images

---

[10] Within a single London borough barriers were rising: 'As the gap between them grows, something fundamental happens to insiders' sense of their place in society. The outsiders become to look less and less like the kind of people insiders mix with at work and socially. They become less recognizable as members of the same society, with a similar right to claim a decent standard of living. At best they are an unsettling embarrassment to be treated with charity. At worst they are an unwanted burden ...' (Leadbeater, 1989, p. 51). From being a regular contributor to *Marxism Today*, Charles Leadbeater went on to become an advisor to Tony Blair.

[11] The extent of British inequalities in wealth in 1989 were truly staggering: 'Noble argues that: "About 500 000 people, one per cent of the population, own just over a third of all private wealth in contemporary Britain and receive just over half of all the personal income derived from possession of wealth". Within this stratum the very rich 50 000, 0.1 per cent of the population, are the most important group' (Scott, 1989, p. 74). In hindsight, however, they were nearly at a historic low point. An inequality in wealth is described here by John Scott that has since soared so that 'staggering' is now too small a word to describe it. John is now PVC (Research) University of Plymouth.

[12] To appreciate such huge differences requires more than words alone can provide: 'Britain is a deeply divided society, and the deepest division of all is the inequality in the ownership of wealth. That the inequalities have persisted for so long helps in itself to legitimate them, to make them more acceptable; the status quo is an influential public relations officer for the rich. And the very extremities of wealth inequalities somehow deprive the statistics of credibility or meaning' (Pond, 1989, p. 189). Chris Pond entered parliament as an MP in 1997 and left it in 2005.

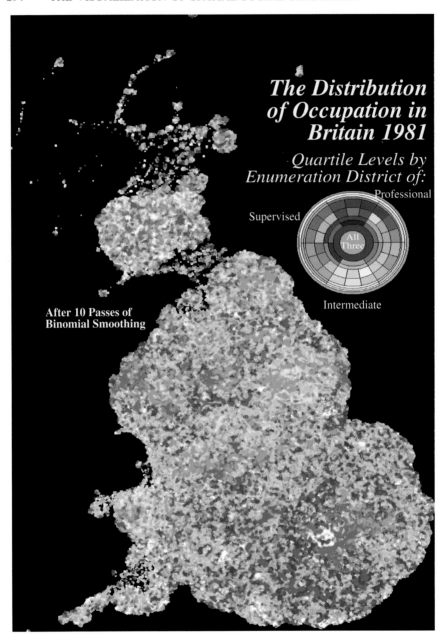

Figure 10.7   Three colour mixing of 129 211 census enumeration districts. The image is possibly made clearer by smoothing, but in places the blue and yellow merge to produce a green, which does not reflect reality. Showing statistics just as they are is often better. Central London is too divided for smoothing to work at all, and hence its white inner core where the dark-red and bright-blue areas clashed.

can open up the world and depict the extent, imbalance and order with both accuracy and in a more emotive way. The images of the world we are shown are the foundations of our understanding. To change it we must know it.[13] To know it we must see it. To see it we must draw it. To draw it we must visualize. Only then can you begin to think from paper to people, from map to mechanism, from screen to society:

> *There were times when the coloured areas on a colour ink-jet map were suddenly obscured by white summer clouds which seemed to scud in from nowhere between the map and the author's eyes, and among which he could glimpse the sparkle of the sea by the coast, the rivers which rolled down to meet it, the towns and villages and people. Sometimes old people materialised out of the map of Norrland and observed with melancholy the exodus of the young towards the coast and the south. From the diagrams which display households suddenly appeared a throng of people who with muted voices told of their lives, of their loneliness, of their joy in their children and of their hopes on their behalf.*
>
> *(Szegö, 1984, p. 30)*

---

[13] We need to see more clearly the social structure we are trying to alter: 'It may seem that the present world is not worth knowing – only worth changing. But to change it one must know it' (Warntz, 1975, p. 75).

Endnote

Finally thanks are due to the British Broadcasting Corporation. Thirty years ago they commissioned the building of the BBC Microcomputer to support the *BBC Computer Literacy Project*. Designed for school children, the manuals for this computer started with over 500 pages devoted to computer programming (from scratch) using BBC BASIC. There were less than 50 pages at the very end of the second volume on how to use the computer as a word processor. BBC BASIC was a sophisticated, structured interpretive programming language that could create coloured graphics using simple commands, allowed recursive functions (ones that can call themselves) and much else.

The series of BBC computer models ended with the BBC A3000, released in 1989. By this time the interface to the machines included multitasking windows, icons and a pointer and menus controlled by a three-buttoned mouse. There were other built-in applications as well as BBC BASIC available on every machine. The Acorn manufacturers included a 'Draw' package to allow the creation, manipulation and visualization of vector graphics. 'Paint' was the equivalent for raster graphics. 'Edit' was a text and binary-file editor with which you could examine any file, such as those created by Draw, or produce Comma-Separated Value (CSV) files. All the graphics in this book were produced on that computer with just those four standard programs.

Only three copies of the original PhD thesis that formed the basis of this work were ever produced. One of the main hopes with revising that text considerably and creating this book is to try to produce a work that could help inspire the next generation of cartographic innovators. Working towards this aim I have included images that do not work very well, as well as ones that do, not all of which got into the original thesis. Most of the graphics in this book have never been printed before, despite being drawn over two decades ago with a view to publication then.

In the seventeenth century Robert Hook looked through his microscope and saw images he knew no one had seen before. Since the 1960s, space images have created a similar sense of excitement for many people. When I first drew some of the images included in this book I thought I was seeing images of our society

The Visualization of Spatial Social Structure, First Edition. Daniel Dorling.
© 2012 John Wiley & Sons, Ltd. Published 2012 by John Wiley & Sons, Ltd.

that no one had seen before, because projections at these scales were then new and the quantity of information unprecedented. Finding no spatial pattern was often as exciting as finding one. However, as the quotations that run through this work have shown, many people had been guessing correctly at much that could now be seen. I did not include quotations illustrating how many others might have changed their views had they been able at the time to have seen some of these images.

After completing the PhD this book is based on I published a few papers on some of the techniques, but moved on to more conventional mapping and analysis, as it proved very hard to then publish high quality colour graphics and as other matters appeared to be becoming more important (rising homelessness, poverty, inequality and injustice). So this work mostly sat on a shelf.

Over 300 floppy discs were filled with back-up files of the work for the thesis when it was in progress. These had been hoarded. With a second-hand Archimedes computer it was possible to find most of the original computer files for the graphics. To be printed here they have been converted to contemporary PDF files. Even in that format, most of these files are unbelievably small by today's standards, largely due to the use of vector graphics. This had one unexpected benefit. The prints could now be produced in this book at a much higher resolution than when originally printed, but unfortunately without the exact colours the original Calcomp printer produced.

One thing that became apparent on seeing these PDFs was how much information could have been put within each cartogram, if they were to be viewed on a high quality screen. No printing would be too small to read. A click of the zoom button could reveal all. Every tiny circle could have had a place name and any other information you wished with insignificant computer/phone/tablet memory requirements.

All the sources here date from over 20 years ago. The images in this book could all be updated to illustrate how Britain changed by very local area from 1990 to 2012. There are three censuses' worth of information to be added to this story, five general elections and several entire new cohort studies have been begun by medical and social scientists since 1981. There is thus more information than ever that is in need of visualizing – and that is just for Britain.

Early on in this book I used an image of the Mandelbrot set, something else seen as very new in the 1980s. Had I revealed the fractal patterns of society? Not quite. For that you possibly have to go down to the level of individual households. Most of us know someone in our street whose social situation is very different from our own, although that has become less common than was the case 30 years ago – less fractal, but more fractured.

Where you live, which part of the town, which side of the street, the river or the railway line, has become increasingly important in affecting what is likely to happen to you. The pictures shown here may look like the portrait of a divided island, but it has become far more divided in the decades since they were drawn.

Acknowledgements

The ideas in this book are almost all drawn from a thesis that was based on two years of research funded by the Strang Studentship of the University of Newcastle upon Tyne. During the course of that research I was aided by many individuals.

The interlibrary loan staff of the Robinson Library kindly waived their normal restrictions. I was allowed to borrow, for free, over a thousand books and papers from around the world in a way almost no postgraduate student would be allowed today.

Judith Houston helped secure the funding, dealt extremely efficiently with the administrative side of my work throughout the period and was always very kind. Many of the staff of the Centre for Urban and Regional Development Studies and the Geography and Planning Departments showed an interest and encouraged me in my work, including Alan Gillard, Tony Champion, Peter Taylor, David Sear and James Cornford.

Colin Wymer, Simon Raybould and Mike Coombes helped satisfy my appetite for digital information from census flows from (what was then) the National On-line Manpower Information System and (what were then) Building Society records.

The researchers associated with the NorthEast Regional Research Laboratory were particularly supportive. Zhilin Li allowed me to access a digital terrain model of part of Scotland. Anna Cross assisted in accessing the Cancer Registry information. Steve Carver provided records of road, rail and land-use data. Chris Brunsdon advised on the analysis of the house price sample.

Stan Openshaw supervised the project, financed much of the equipment and supplied the local election and 1971 Census information. He also gave me the space and encouragement I needed and, most importantly, Stan could see little wrong in someone doing something unusual, giving me the freedom to think and work, which today's postgraduate student is unlikely to encounter now that theses are so often produced to a specific recipe, so that chapter titles are mostly preassigned.

Thanks are especially due to Martin Charlton, who read the original document that became the basis for this book and who spent many hours helping me amass

The Visualization of Spatial Social Structure, First Edition. Daniel Dorling.
© 2012 John Wiley & Sons, Ltd. Published 2012 by John Wiley & Sons, Ltd.

the vast majority of the information used here, as well as permitting extensive use of a great deal of expensive equipment. Bruce Tether spent many days assisting with the editing and collection of the thousands of election results used, and gave useful criticism and advice. Richard Park read and commented on the final draft of the original thesis. Ile Ashcroft and Edward Jones corrected much of the English, while Stacy Hewitt gave the work a professional proof-reading.

Eric Charlesworth advised on the style at an early stage, as well as providing geographical advice. Bronwen Dorling meticulously corrected my writing and gave constant encouragement (she had originally taught me to read 12 years earlier, at age 9, and has – from 1977 until 2012 – had to deal with my attempts at writing). David Dorling helped rearrange many of the ideas presented here and first taught me to program. He also recovered all the graphics from the originally floppy disk back-ups so that they could be included here (and seen printed properly for the first time). Finally Anna Macdonald spent several weeks referencing small scale maps and typing in numerous extensive quotations and tables of data. She also had to put up with my obsession to finish the original PhD thesis a year early. Nothing is achieved in isolation.

Apart from a little tweaking these words immediately above were written over twenty years ago. The words further above, in the rest of this book, have been altered far more. Since 1991 the thesis this book is based on has been gathering dust on a shelf in the Robinson Library at the University of Newcastle upon Tyne. A copy of the text and low quality scans of the graphics were put on the web, but they are hard to find and difficult to appreciate – taken out of book form. It is Debbie Jupe, Commissioning Editor of Statistics and Mathematics books for Wiley-Blackwell, whom I have to thank for sending an email in June 2011 asking if I might be interested in working the thesis up into a book. She and five extremely positive anonymous reviewers and one more restrained (but not completely dismissive) one are responsible for this work being printed twenty years after it was first written. Richard Davies, the Assistant Production Manager at the Chichester offices of Wiley was also extremely helpful, as were Martin Dodge of Manchester University, Scott Orford and Jonathan Radcliffe of Cardiff University and Mark Green of Sheffield University all very kindly commented on drafts and further revisions to the original text. I am also particularly grateful to Patricia Bateson who proof read the final text, Prachi Sinha Sahay in Singapore who oversaw final production and Sangeetha Parthasarathy in Chennai who typeset the work so diligently.

Finally, I should also thank the examiners of that original thesis, Peter Haggett of Bristol University and Peter Taylor of Newcastle University. They were also very kind, ignored my spelling mistakes and my cavalier attitude to conventional ways of ordering a thesis to examine instead what it was I was actually discovering, not whether I was jumping through hoops in the correct way to find it. I ended up teaching in Bristol the courses Peter Haggett had taught, and what I studied moved, in the intervening years, more towards what Peter Taylor thought mattered. I have tried to use their comments to make this a little more relevant than the original. I have also cut out a lot of text that seemed to be very important at the time, but in hindsight, as with so much in life, wasn't.

Appendix: Drawing faces

```
REM *****************************************
REM ** BBC BASIC procedure to draw a face **
REM *****************************************
DEF
  PROCdraw_face(Draw%,x%,y%,r%,Fill_colour%,
  Line_colour%,Line_width%,cheeks,eyes,nose,mouth,
  RETURN face_boundary%())
LOCAL Word%, Movepath%, Bezierpath%, Closepath%
LOCAL Line_style%, Points%, block%
Word% = 4
Movepath% = 2
Bezierpath% = 6
Closepath% = 5
Points% = 97*Word%
Line_style% = FNdraw_line_style(1,1,1,1,0,0)
REM *Simple function describing a basic line*
r% = r%/2.5
DIM block% Points%
block%!0 = Movepath%
block%!Word% = x%-r%*2
block%!(Word%*2) = y%+r%*2
block%!(Word%*3) = Bezierpath%
block%!(Word%*4) = x%-r%*1.3
block%!(Word%*5) = y%+r%*3.13
block%!(Word%*6) = x%+r%*1.3
block%!(Word%*7) = y%+r%*3.13
block%!(Word%*8) = x%+r%*2
block%!(Word%*9) = y%+r%*2
block%!(Word%*10) = Bezierpath%
block%!(Word%*11) = x%+r%*(2.78+cheeks*2)
block%!(Word%*12) = y%+r%*0.73
block%!(Word%*13) = x%+r%*1.49
block%!(Word%*14) = y%-r%*2
block%!(Word%*15) = x%
```

The Visualization of Spatial Social Structure, First Edition. Daniel Dorling.
© 2012 John Wiley & Sons, Ltd. Published 2012 by John Wiley & Sons, Ltd.

```
block%!(Word%*16) = y%-r%*2
block%!(Word%*17) = Bezierpath%
block%!(Word%*18) = x%-r%*1.49
block%!(Word%*19) = y%-r%*2
block%!(Word%*20) = x%-r%*(2.78+cheeks*2)
block%!(Word%*21) = y%+r%*0.73
block%!(Word%*22) = x%-r%*2
block%!(Word%*23) = y%+r%*2
block%!(Word%*24) = Closepath%
REM left eye
block%!(Word%*25) = Movepath%
block%!(Word%*26) = x%-r%*(7+3*eyes)/8
block%!(Word%*27) = y%+r%*(10-3*eyes)/8
block%!(Word%*28) = Bezierpath%
block%!(Word%*29) = x%-r%*(7+3*eyes)/8
block%!(Word%*30) = y%+r%*(10-3*eyes)/8
block%!(Word%*31) = x%-r%*(5+eyes)/8
block%!(Word%*32) = y%+r%*(9-4*eyes)/8
block%!(Word%*33) = x%-r%/2
block%!(Word%*34) = y%+r%*(10-3*eyes)/8
block%!(Word%*35) = Bezierpath%
block%!(Word%*36) = x%+r%*(2+2*eyes)/8
block%!(Word%*37) = y%+r%*(16-eyes)/8
block%!(Word%*38) = x%-r%*(13+5*eyes)/8
block%!(Word%*39) = y%+r%*(16-eyes)/8
block%!(Word%*40) = x%-r%*(7+3*eyes)/8
block%!(Word%*41) = y%+r%*(10-3*eyes)/8
block%!(Word%*42) = Closepath%
REM right eye
block%!(Word%*43) = Movepath%
block%!(Word%*44) = x%+r%*(7+3*eyes)/8
block%!(Word%*45) = y%+r%*(10-3*eyes)/8
block%!(Word%*46) = Bezierpath%
block%!(Word%*47) = x%+r%*(7+3*eyes)/8
block%!(Word%*48) = y%+r%*(10-3*eyes)/8
block%!(Word%*49) = x%+r%*(5+eyes)/8
block%!(Word%*50) = y%+r%*(9-4*eyes)/8
block%!(Word%*51) = x%+r%/2
block%!(Word%*52) = y%+r%*(10-3*eyes)/8
block%!(Word%*53) = Bezierpath%
block%!(Word%*54) = x%-r%*(2+2*eyes)/8
block%!(Word%*55) = y%+r%*(16-eyes)/8
block%!(Word%*56) = x%+r%*(13+5*eyes)/8
block%!(Word%*57) = y%+r%*(16-eyes)/8
block%!(Word%*58) = x%+r%*(7+3*eyes)/8
block%!(Word%*59) = y%+r%*(10-3*eyes)/8
block%!(Word%*60) = Closepath%
REM nose
block%!(Word%*61) = Movepath%
```

```
block%!(Word%*62) = x%
block%!(Word%*63) = y%+r%
block%!(Word%*64) = Bezierpath%
block%!(Word%*65) = x%
block%!(Word%*66) = y%+r%
block%!(Word%*67) = x%-r%*(1.1+nose*2/3)
block%!(Word%*68) = y%
block%!(Word%*69) = x%
block%!(Word%*70) = y%
block%!(Word%*71) = Bezierpath%
block%!(Word%*72) = x%+r%*(1.1+nose*2/3)
block%!(Word%*73) = y%
block%!(Word%*74) = x%
block%!(Word%*75) = y%+r%
block%!(Word%*76) = x%
block%!(Word%*77) = y%+r%
block%!(Word%*78) = Closepath%
REM mouth
block%!(Word%*79) = Movepath%
block%!(Word%*80) = x%-r%*0.8
block%!(Word%*81) = y%+r%*(mouth/4-0.75)
block%!(Word%*82) = Bezierpath%
block%!(Word%*83) = x%-r%*0.8
block%!(Word%*84) = y%+r%*(mouth/4-0.75)
block%!(Word%*85) = x%
block%!(Word%*86) = y%-r%*(8+mouth*6)/8
block%!(Word%*87) = x%+r%*0.8
block%!(Word%*88) = y%+r%*(mouth/4-0.75)
block%!(Word%*89) = Bezierpath%
block%!(Word%*90) = x%+r%*0.8*(5-mouth)/8
block%!(Word%*91) = y%-r%*(9+mouth*7)/8
block%!(Word%*92) = x%-r%*0.8/2
block%!(Word%*93) = y%-r%*(8+mouth*8)/8
block%!(Word%*94) = x%-r%*0.8
block%!(Word%*95) = y%+r%*(mouth/4-0.75)
block%!(Word%*96) = Closepath%
PROCdraw_path_object(Draw%,Fill_colour%,Line_colour%,
  Line_width%,Line_style%,block%,Points%,face_boundary%())
ENDPROC
REM ****************************************
DEF
  PROCdraw_path_object(Draw%,Fill_colour%,Line_colour%,
  Line_width%,Line_style%,Pointer%,Points%,RETURN
|parent_boundary%())
LOCAL pointer%, Path%, Endpath%, tag%, Word%,
  boundary%(), point%
Path% = 2
Endpath% = 0
Word% = 4
```

```
DIM boundary%(4)
boundary%()=0,2^30,2^30,-(2^30),-(2^30)
pointer% = FNdraw_start_object(Path%)
PROCdraw_word(Fill_colour%)
PROCdraw_word(Line_colour%)
PROCdraw_word(Line_width%)
PROCdraw_word(Line_style%)
point%=0
REPEAT
tag%=Pointer%!point%
PROCdraw_word(tag%)
point% += Word%
CASE tag% OF
WHEN 2,8
PROCdraw_coords(Pointer%!point%,Pointer%!(point%+Word%),
  boundary%())
point% += 2*Word%
WHEN 6
PROCdraw_coords(Pointer%!point%,Pointer%!(point%+Word%),
  boundary%())
point% += 2*Word%
PROCdraw_coords(Pointer%!point%,Pointer%!(point%+Word%),
  boundary%())
point% += 2*Word%
PROCdraw_coords(Pointer%!point%,Pointer%!(point%+Word%),
  boundary%())
point% += 2*Word%
ENDCASE
UNTIL point% = Points%
PROCdraw_word(Endpath%)
PROCdraw_end_object(pointer%, boundary%())
PROCdraw_extend_boundary(parent_boundary%(), boundary%())
ENDPROC
```

References

Agnew J.A. (1987) *Place and Politics: The Geographical Mediation of State and Society*, Allen and Unwin, Boston, Massachusetts.

Anderson J.M. (1988) Human cartography: mapping the world of man (review), *Cartographica*, **25** (4), 88–89.

Anderson G.C. (1989) Images worth thousands of bits of data, *The Scientist*, **3** (3), 1, 16–17.

Angel S. and Hyman G.M. (1972) Transformations and geographic theory, *Geographical Analysis*, **4**, 350–367.

Angel S. and Hyman G.M. (1976) Urban Fields: A Geometry of Movement for Regional Science, Pion Limited, London.

Applied Urbanetics INC (1971) Reading the map: social characteristics of Washington DC, *Metropolitan Bulletin*, **6**, 4–5.

Arnheim R. (1970) *Visual Thinking*, Faber and Faber Ltd, London.

Arnheim R. (1976) The perception of maps, *The American Cartographer*, **3** (1), 5–10.

Bachi R. (1968) *Graphical Rational Patterns: A New Approach to Graphical Presentation of Statistics*, Israel Universities Press, Jerusalem.

Ballard B. and Norris P. (1983) User needs – an overview, in *A Census User's Handbook* (ed. D. Rhind), Methuen, London.

Bashshur R.L., Shannon G.W. and Metzner C.A. (1970) The application of three-dimensional analogue models to the distribution of medical care facilities, *Medical Care*, **8** (5), 395–407.

Becker R.A., Cleveland W.S. and Weil G. (1988) The use of brushing and rotation for data analysis, in *Dynamic Graphics for Statistics* (eds W.S. Cleveland and M.E. McGill), Wandsworth, California.

Becker R.A., Cleveland W.S. and Wilks A.R. (1988) Dynamic graphics for data analysis, in *Dynamic Graphics for Statistics* (eds W.S. Cleveland and M.E. McGill), Wandsworth, California.

Becker R.A., Eick S.G., Miller E.O. and Wilks A.R. (1990a) Network visualization, *Proceedings of the 4th International Symposium on Spatial Data Handling*, Zurich, vol. 1, pp. 285–294.

Becker R.A., Eick S.G., Miller E O. and Wilks A.R. (1990b) Dynamic graphics for network visualization, *Visualization '90, Proceedings of the First IEEE Conference on Visualization*, October, San Francisco, California, pp. 93–96.

Begg I. and Eversley D. (1986) Deprivation in the inner city: social indicators from the 1981 census, in *Critical Issues in Urban Economic Development* (ed. V.A. Hausner), Chapter 3, Clarendon Press, Oxford.

Bergeron R.D. and Grinstein G.G. (1989) A reference model for the visualization of multi-dimensional data, in *Eurographics '89*, Elsevier Science Publishers.

Bertin J. (1981) Graphics and Graphic Information – Processing, Walter de Gruyter, Berlin.

Bertin J. (1983) *Semiology of Graphics* (translation of *Semiologie Graphique* with W.J. Berg as translator), The University of Wisconsin Press, Madison, Wisconsin.

Bickmore D.P. (1975) The relevance of cartography, in *Display and Analysis of Spatial Data* (eds J.C. Davies and M.J. McCullagh), John Wiley & Sons, Ltd, London, pp. 328–351.

Blot W.J. and Fraumeni J.F. (1982) Geographical epidemiology of cancer in the United States, in *Cancer Epidemiology and Prevention* (eds D. Schutter and J. Fraumeni), W.B. Sanders & Co., Philadelphia.

Bochel J.M. and Denver D.T. (1988) Scottish district elections 1988: results and statistics, Electoral Studies, University of Dundee.

Bochel J.M. and Denver D.T. (1990) The Scottish regional elections 1990: results and statistics, Electoral Studies, University of Dundee.

Borchert J.R. (1987) Maps, geography, and geographers, *The Professional Geographer*, **39** (4), 387–389.

Brannon G. (1989) The artistry and science of map-making, *The Geographical Magazine*, **61** (9), 38–40.

Brant J. (1984) Patterns of migration from the 1981 census, *Population Trends*, **35**, 23–30.

Braudel F. (1979) *The Wheels of Commerce*, Fontana Press, London.

Buck N., Gordon I., Young K., Ermish J. and Mills L. (1986) *The London Employment Problem, Economic and Social Research Council, Inner Cities Research Programme*, Clarendon Press, Oxford.

Budge I. and Farlie D.J. (1983) *Explaining and Predicting Elections: Issue Effects and Party Strategies in Twenty-Three Democracies*, George Allen and Unwin, London.

Bulusu L. (1990) Internal migration in the UK, 1989, *Population Trends*, **62**, 3–36.

Bunge W.W. (1964) Patterns of location, Discussion Paper No. 3, Michigan Inter-University Community of Mathematical Geographers, University of Michigan, Ann Arbor.

Bunge W.W. (1966) Theoretical geography, Lund Studies in Geography, Series C, No. 1, pp. 1–285.

Bunge W.W. (1968) Spatial prediction, in *The Philosophy of Maps* (ed. J.D. Nystuen), Discussion Paper No.12, Michigan Inter-University Community of Mathematical Geographers, Wayne State University, pp. 31–33.

Bunge W.W. (1973) The geography of human survival, *Annals of the Association of American Geographers*, **63** (3), 275–295.

Bunge W.W. (1975) Detroit humanly viewed. the American urban present, in *Human Geography in a Shrinking World* (eds R. Abler, D. Janelle, A. Philbrick and J. Sommer), Duxbury Press, Massachusetts.

Bunge, W. (2011) *Fitzgerald: Geography of a Revolution (Geographies of Justice and Social Transformation)*, University of Georgia Press (reprinted from 1971 original: http://en.wikipedia.org/wiki/William_Bunge).

Burnham W.D. (1978) Great Britain: the death of the collectivist consensus?, in *Political Parties: Development and Decay* (eds L. Maisel and J. Coopwe), Sage, London.

Carlstein T. (1982) *Time Resources, Society and Ecology*, George Allen and Unwin, London.

Cauvin C., Schneider C. and Cherrier G. (1989) Cartographic transformations and the piezopleth maps method, *The Cartographic Journal*, **26**, 96–104.

Champion A.G. (1989) Internal migration and the spatial distribution of population, in *The Changing Population of Britain* (ed. H. Joshi), Basil Blackwell, Oxford.

Champion A.G. and Green A. (1989) Local economic differentials and the 'north–south divide', in *The North–South Divide: Regional Change in Britain in the 1980s* (eds J. Lewis and A. Townsend), Paul Chapman, London.

Champion A.G., Green A.E., Owen D.W., Ellin D.J. and Coombes M.G. (1987) *Changing Places: Britain's Demographic, Economic and Social Complexion*, Edward Arnold, London.

Chernoff H. (1973) The use of faces to represent points in k-dimensional space graphically, *Journal of the American Statistical Association*, **68** (342), 361–368.

Cliff A.D. and Haggett P. (1988) *Atlas of Disease Distributions: Analytic Approaches to Epidemiological Data*, Basil Blackwell Ltd, Oxford.

Congdon P. (1989) An analysis of population and social change in London wards in the 1980s, *Transactions of the Institute of British Geographers*, **14**, 478–491.

Congdon P. and Champion A. (1989) Trends and structure in London's migration and their relation to employment and housing markets, in *Advances in Regional Demography: Information, Forecasts, Models* (eds P. Congdon and P. Batey), Belhaven Press, London.

Coombes M.G., Dixon J.S., Goddard J.B., Openshaw S. and Taylor P.J. (1978) Towards a more rational consideration of census areal units: daily urban systems in Britain, *Environment and Planning A*, **10**, 1179–1185.

Craig J. (1981) Migration patterns of Surrey, Devon and South Yorkshire, *Population Trends*, **23**, 16–21.

Craig J. (1987) Changes in the population composition of England and Wales since 1841, *Population Trends*, **48**, 27–36.

Craig J. (1988) Population density and concentration in England and Wales 1971 and 1981, Studies on Medical and Population Subjects No. 52, Office of Population Censuses and Surveys.

Crewe I. (1988) Voting patterns since 1959, *Contemporary Record*, **2** (4), 2–6.

Crewe I., Särlvik B. and Alt J. (1977) Partisan dealignment in Britain 1964–1974, *British Journal of Political Science*, **7**, 129–190.

Cuff D.J. (1989) Human cartography (review), *The American Cartographer*, **16** (1), 52–53.

Cuff D.J., Pawling J.W. and Blair E.T. (1984) Nested value-by-area cartograms for symbolizing land use and other proportions, *Cartographica*, **21** (4), 1–8.

Curtice, J. and Steed M. (1988) Analysis, in *The British General Election of 1987* (eds D. Butler and D. Kavanagh), Macmillan, London.

Dale L. (1971) Cartographic representation of journey-to-work movements – 1971 Canadian census, Working Paper (Demographic and Socio-economic Series) No. 8, Statistics Canada, Ottawa.

Davis N. and Walker C. (1975) Migrants entering and leaving the UK, 1964–74, *Population Trends*, **1**, 2–5.

DeFanti T.A., Brown M.D. and McCormick B.H. (1989) Visualization: expanding scientific and engineering research opportunities, *Computer*, **22** (8), 12–25.

della Dora, V. (2011) *Imagining Mount Athos: Visions of a Holy Place from Homer to World War II*, University of Virginia Press, Charlottesville.

Denes A. (1979) *Isometric Systems in Isotropic Space: Map Projections: From the Study of Distortions Series 1973–1979*, Visual Studies Workshop Press, New York.

Denver D. (1989) *Elections and Voting Behaviour in Britain*, Philip Allan, New York.

Dickinson R.R. (1989) Interactive 4-D visualization of fields, Technical Report CS-89-5, Department of Computer Science, University of Waterloo, Ontario, Canada.

Donoho D.L., Huber P.J., Ramos E. and Thoma H.M. (1988) Kinematic display of multivariate data, in *Dynamic Graphics for Statistics* (eds W.S. Cleveland and M.E. McGill), Wandsworth, California.

Dooley D. and Cohen M.F. (1990) Automatic illustration of 3D geometric models: surfaces, *Visualization '90, Proceedings of the First IEEE Conference on Visualization*, October, San Francisco, California, pp. 307–314.

Dorling D. (1994) Bringing elections back to life, *Geographical Magazine*, **66** (12), 20–21.

Dorling D. (1996) Area cartograms: their use and creation, Concepts and Techniques in Modern Geography Series No. 59, University of East Anglia, Environmental Publications.

Dorling D. (2011) *So You Think You Know about Britain?*, Constable, London.

Dorling D., Barford A. and Newman M. (2006) WORLDMAPPER: the world as you've never seen it before, *IEEE Transactions on Visualization and Computer Graphics*, **12** (5), 757–764.

Dorling D. and Thomas B. (2011) *Bankrupt Britain: An Atlas of Social Change*, Policy Press, Bristol.

Dougenik J.A., Chrisman N.R. and Niemeyer D.R. (1985) An algorithm to construct continuous area cartograms, *Professional Geographer*, 37 (1), 75–81.

Dougenik J.A., Niemeyer D.R. and Chrisman N.R. (1983) A computer algorithm to build continuous area cartograms, Harvard Computer Graphics Week, Harvard University Graduate School.

Duncan O.D., Cuzzort R.P. and Duncan B. (1961) *Statistical Geography: Problems in Analyzing Areal Data*, The Free Press of Glencoe, Illinois.

Dunleavy P. (1983) Voting and the electorate, in *Developments in British Politics* (ed. H. Drucker), Macmillan, London.

Eastman J.R., Nelson W. and Shields G. (1981) Production considerations in isodensity mapping, *Cartographica*, **18** (1), 24–30.

Elliott, A. (2011) Committee on Medical Aspects of Radiation in the Environment (COMARE): Fourteenth Report, Health Protection Agency, London.

Evans D. (1973) Keynote Address, Why is Computer Graphics Always a Year Away?, *Computer Graphics, Quarterly Report of Siggraph-ACM*, **8** (1), 5–11.

Foley J.D., van Dam A., Feiner S.K. and Hughes J.F. (1990) *Computer Graphics: Principles and Practise*, Addison-Wesley, Reading.

Forbes J. (1984) Problems of cartographic representation of patterns of population change, *The Cartographic Journal*, **22**, 93–102.

Fothergill S. and Gudgin G. (1982) *Unequal Growth: Urban and Regional Change in the UK*, Heinemann Books, London.

Freeman S., (1991) Envisioning information: review, *IEEE Computer Graphics and Applications*, **11** (1), 113–114.

Frenkel K.A. (1988) The art and science of visualizing data, *Communications of the ACM*, **31** (2), 110–121.

Frost M.E. and Spence N.A. (1981) Unemployment, structural economic change and public policy in British regions, in *Progress in Planning*, vol. 16, Pergamon Press, Oxford, pp. 1–130.

Goddard J.B. (1983) Structural change in the British space economy, in *The Urban and Regional Transformation of Britain* (eds J.B. Goddard and A.G. Champion), Methuen, London.

Goodchild M.F. (1988) Stepping over the line: technological constraints and the new cartography, *The American Cartographer*, **15** (3), 311–319.

Gudgin G. and Taylor P.J. (1973) Electoral bias and the distribution of party voters, Seminar Papers, No. 22, Department of Geography, Newcastle University.

Gyford J., Leach S. and Game C. (1989) *The Changing Politics of Local Government*, Unwin Hyman, London.

Hagen C.B. (1982) Maps: an overview of the producer–user interaction, *ACSM-ASP 42nd Annual Convention*, March, Denver, Colorado, pp. 325–338.

Haggerty M. (1991) The art of artificial reality, *IEEE Computer Graphics and Applications*, **11** (1), 8–14.

Halsey A.H. (1989) Social trends since World War II, in *Divided Nation: Social and Cultural Change in Britain* (eds L. McDowell, P. Sarre and C. Hamnett), Hodder and Stoughton and The Open University, London.

Hamnett C. (1986) The changing socio-economic structure of London and the South East, 1961–81, *Regional Studies*, **20** (5), 391–406.

Hamnett C. (1987) A tale of two cities: sociotenurial polarisation in London and the South-East, 1966–81, *Environment and Planning A*, **19**, 537–556.

Hamnett C. (1989) The owner-occupied housing market in Britain: a north–south divide, in *The North–South Divide: Regional Change in Britain in the 1980s* (eds J. Lewis and A. Townsend), Paul Chapman, London.

Hamnett C. and Randolph B. (1983) The changing tenure structure of the Greater London housing market, 1961–1981, *The London Journal*, **9** (2), 153–164.

Hardy R.L. (1988) Concepts and results of mapping in three dimensional space, Technical Papers of the ACSM-ASPRS Annual Convention, St Louis, Missouri, vol. 2, pp. 106–115.

Harley J.B. (1988) Maps, knowledge, and power, in *The Iconography of Landscape* (eds D. Cosgrove and S. Daniels), Cambridge University Press, New York.

Harley J.B. (1990) Cartography, ethics and social theory, *Cartographica*, Vol. **27**, No. 2, pp. 1–27.

Harvey D. (1990) Between space and time: reflections on the geographical imagination, *Annals of the Association of American Geographers*, **80** (3), 418–434.

Hennig, B.D., Pritchard, J., Ramsden M. and Dorling, D. (2010) Remapping the world's population: visualizing data using cartograms, reprinted from *ArcUser Magazine*, No. 1, as Chapter 51 of Dorling D. (ed.) (2010), *Fair Play*, Policy Press, Bristol, pp. 66–69.

Herr L. (1990) Volume visualization for biology – a new dimension in computer graphics, *Bioscience*, **40** (3), 199–202.

Herzog A. (1989) Modelling reliability on statistical surfaces by polygon filtering, in *The Accuracy of Spatial Databases* (eds M. Goodchild and S. Gopal), Chapter 18, Taylor & Francis, New York.

Hollingsworth T.H. (1970) Migration: a study based on Scottish experience between 1939 and 1964, Social and Economic Studies, Glasgow University, Occasional Paper No. 12, Oliver & Boyd.

Holly B.P. (1978) The problem of scale in time-space research, in *Time and Regional Dynamics* (eds T. Carlstein, D. Parkes and N. Thrift), Edward Arnold, London.

Huggins W. (1973) What's needed now?, Why is computer graphics always a year away?, *Computer Graphics, Quarterly Report of Siggraph-ACM*, **8** (1), 32–48.

Hunt A.J. (1968) *Problems of Population Mapping: An Introduction*, Institute of British Geographers, Transactions No. 43, April 1968, George Philip & Son, London.

Hunter J.M. and Meade M.S. (1971) Population models in the high school, *The Journal of Geography*, **70**, 95–104.

Hunter J.M. and Young J.C. (1968) A technique for the construction of quantitative cartograms by physical accretion models, *The Professional Geographer*, **20** (6), 402–407.

Jacob R.J.K., Egeth H.E. and Bevin W. (1976) The face as a data display, *Human Factors*, **18**(2), 189–200.

Jobse R.B. and Musterd S. (1989) Changes in migration within the Netherlands, 1975–85, *Tijdschrift voor Economische en Sociale Geografie*, **80**, 244–250.

Johnson J.H. (1984) Inter-urban migration in Britain: a geographical perspective, in *Migration and Mobility: Biosocial Aspects of Human Movement* (ed. A.J. Boyce), Taylor & Francis, London.

Johnson J.H., Salt J. and Wood P.A. (1974) *Housing and the Migration of Labour in England and Wales*, Saxon House/Lexington Books, England.

Johnston R.J. (1982) Short-term electoral change in England: estimates of its spatial variation, *Political Geography Quarterly*, **1**(1), 41–55.

Johnston R.J. and Pattie C.J. (1988) Changing voter allegiances in Great Britain, 1979–1987: an exploration of regional patterns, *Regional Studies*, **22** (3), 179–192.

Johnston R.J. and Pattie C.J. (1989a) Voting in Britain since 1979: a growing north–south divide, in *The North–South Divide: Regional Change in Britain in the 1980s* (eds J. Lewis and A. Chapman), Paul Chapman, London.

Johnston R.J. and Pattie C.J. (1989b) A growing north–south divide in British voting patterns, 1979–87, *Geoforum*, **20** (1), 93–106.

Kadmon N. and Shlomi E. (1978) A polyfocal projection for statistical surfaces, *The Cartographic Journal*, **15** (1), 36–41.

Ke Y. and Panduranga E.S. (1990) A journey into the fourth dimension, *Visualization '90, Proceedings of the First IEEE Conference on Visualization*, October, San Francisco, California, pp. 219–229.

Kelly J. (1987) Constructing an area-value cartogram for New Zealand's population, *New Zealand Cartographer*, **17** (1), 3–10.

Kerlick G.D. (1990) Moving iconic objects in scientific visualization, *Visualization '90, Proceedings of the First IEEE Conference on Visualization*, October, San Francisco, California, pp. 124–130.

King R. and Shuttleworth I. (1989) The Irish in Coventry: the social geography of a relict community, *Irish Geography*, **22** (2), 64–78.

Kleiner B. and Hartigan J.A. (1981) Representing points in many dimensions by trees and castles, *Journal of the American Statistical Association*, **76** (374), 260–276.

Knox G. (1964) Epidemiology of childhood leukaemia in Northumberland and Durham, *British Journal of Preventative and Social Medicine*, **18**, 17–24.

Kosslyn S.M. (1983) *Ghost in the Mind's Machine: Creating and Using Images in the Brain*, W.W. Norton & Company, New York.

Kraak M.J. (1989) Computer-assisted cartographical 3D imaging techniques, in *GIS: Three Dimensional Applications in Geographic Information Systems* (ed. J. Raper), Taylor & Francis, London.

Krueger M.W. (1983) Artificial Reality, Addison-Wesley, Reading, Massachusetts.

La Breque M. (1989) The scientific uses of visualization, *American Scientist*, **77** (6), 525–527.

Lai P.C. (1983) A new solution to travel time map transformations, PhD ISBN 0-315-13480-1, The University of Waterloo, Canada.

Lavin S. (1986) Mapping continuous geographical distributions using dot density shading, *The American Cartographer*, **13** (2), 140–150.

Lawton R. (1968) The journey to work in Britain: some trends and problems, *Regional Studies*, **2**, 27–40.

Leadbeater C. (1989) In the land of the dispossessed, in *Divided Nation: Social and Cultural Change in Britain* (eds L. McDowell, P. Sarre and C. Hamnett), Hodder and Stoughton and The Open University, London.

Levkowitz H. (1988) Color in computer graphic representation of two-dimensional parameter distributions, PhD No. DA8824759, University of Pennsylvania.

Lewis P.F. (1969) Impact of negro migration on the electoral geography of Flint, Michigan, 1932–1962, in *The Structure of Political Geography* (eds R.E. Kasperson and J.V. Minghi), University of London Press.

Lewis J. and Townsend A. (eds) (1989) *The North–South Divide: Regional Change in Britain in the 1980s*, Paul Chapman, London.

Long M.B., Lyons K. and Lam J.K. (1990) Acquisition and representation of two- and three- dimensional data from turbulent flow, in *Visualization in Scientific Computing* (eds G.M. Nielson, B. Shriver and L.J. Rosenblum), IEEE Computer Society Press, California, pp. 132–139.

McCleary G.F. (1988) The war atlas: armed conflict – armed peace (review), *Cartographica*, **25** (3), 147–149.

McCormick B.H., DeFanto T.A. and Brown M.D. (eds) (1987) Visualization in scientific computing, *Computer Graphics*, **21** (6).

McDonald J.A. (1988) Orion I: interactive graphics for data analysis, in *Dynamic Graphics for Statistics* (eds W.S. Cleveland and M.E. McGill), Wandsworth, California.

McKee C. (1989) Local level socio-economic and demographic change in the south-east of England, 1971 to 1981, DPhil Dissertation, Geography Department, Birkbeck College, University of London.

Marchand B. (1973) Deformation of a transportation surface, *Annals of the Association of American Geographers*, **63** (4), 507–521.

Marquand J. (1983) The changing distribution of service employment, in *The Urban and Regional Transformation of Britain* (eds J.B. Goddard and A.G. Champion), Methuen, London.

Marr D. (1982) *Vision*, W.H. Freeman & Co., New York.

Massey D.S. and Stephan G.E. (1977) The size-density hypothesis in Great Britain: analysis of a deviant case, *Demography*, **14** (3), 351–361.

Miller W.L. (1990) Voting and the electorate, in *Developments in British Politics* (eds P. Dunleavy, A. Gamble and G. Peele), vol. 3, Macmillan, London.

Miller W.L., Raab G. and Britto K. (1974) Voting research and the population census 1918–71: surrogate data for constituency analyses, *Journal of the Royal Statistical Society A*, **137** (3), 384–411.

Mills M.I. (1981) A study of the human response to pictoral representations in telidon, Telidon Behavioural Research 3, The Behavioural Research and Evaluation Group, Information Branch, Ottawa, Canada.

Mounsey H.M. (1982) Mapping population change through time by computer and cine film, in *Computers in Cartography* (eds D. Rhind and T. Adams), British Cartographic Society, Special Publication No. 2, London, pp. 127–132.

Muehrcke P.C. (1972) Thematic cartography, Resource Paper No.19, Association of American Geographers, Washington.

Muehrcke P.C. (1978) *Map Use: Reading, Analysis, and Interpretation*, JP Publicatins, Madison, Wisconsin.

Muehrcke P.C. (1981) Maps in geography, *Cartographica*, Monograph 27 (ed. L. Guelke), *Maps in Modern Geography – Geographical Perspectives on the New Cartography*, **18** (2), 1–41.

Mueser P. (1989) The spatial structure of migration: an analysis of flows between states in the USA over three decades, *Regional Studies*, **23** (3), 185–200.

Muller J.C. (1989) Changes ahead for the mapping profession, *Proceedings of Auto Carto 9, International Symposium on Computer-Assisted Cartography*, pp. 675–683.

Murphy M.J. (1989) Housing the people: from shortage to surplus?, in *The Changing Population of Britain* (ed. H. Joshi), Basil Blackwell, Oxford.

Nelson B. and McGregor B. (1983) A modification of the non-contiguous area cartogram, *New Zealand Cartographic Journal*, 12 (2), pp. 21–29.

Nielson G.M., Shriver B. and Rosenblum L.J. (eds) (1990) *Visualization in Scientific Computing*, IEEE Computer Society Press, Los Alamitos, California.

Norris P. and Mounsey H.M. (1983) Analysing change through time, in *A Census User's Handbook* (ed. D. Rhind), Methuen, London.

Olson J.M. (1976) Noncontiguous area cartograms, *The Professional Geographer*, **28**, 371–380.

Openshaw S. (1983) Multivariate analysis of census data: the classification of areas, in *A Census User's Handbook* (ed. D. Rhind), Methuen, London.

Papathomas T.V. and Julesz B. (1988) The application of depth separation to the display of large data sets, in *Dynamic Graphics for Statistics* (eds W.S. Cleveland and M.E. McGill), Wandsworth, California.

Parslow R. (1987) A new direction for computer graphics, *Computer Bulletin*, September, 22–25.

Peach C. (1982) The growth and distribution of the black population in Britain 1945–1980, in *Demography of Immigrants and Minority Groups in the UK* (ed. D.A. Coleman), Academic Press, London.

Peddle J.B. (1910) *The Construction of Graphical Charts*, McGraw-Hill Book Company, New York.

Phillips R.J. (1989) Are maps different from other kinds of graphic information?, *The Cartographic Journal*, **26**, 24–25.

Plantinga W.H. (1988) The asp: a continuous, viewer centred object representation for computer vision, PhD No. DA8820061, University of Wisconsin-Madison.

Pond C. (1989) Wealth and the two nations, in *Divided Nation: Social and Cultural Change in Britain* (eds L. McDowell, P. Sarre and C. Hamnett), Hodder and Stoughton and The Open University, London.

Pred A. (1984) Place as historically contingent process: structuration and the time-geography of becoming places, *Annals of the Association of American Geographers*, **74** (2), 279–297.

Pred A. (1986) *Place, Practice and Structure: Social and Spatial Transformation in Southern Sweden: 1750–1850*, Barnes and Noble Books, Totowa, New Jersey.

Rahu M. (1989) Graphical representation of cancer incidence data: Chernoff faces, *International Journal of Epidemiology*, **18** (4), 763–767.

Redford A. (1976), *Labour Migration in England, 1800–1850*, Manchester University Press, Manchester.

Rhind D. (1975a) The reform of areal units, *Area*, **7** (1), 1–3.

Rhind D. (1975b) Geographical analysis and mapping of the 1971 UK census data, Census Research Unit Working Paper No. 3, University of Durham.

Rhind D. (1975c) Mapping the 1971 census by computer, *Population Trends*, **2**, 9–12.

Rhind D. (ed.) (1983) *A Census User's Handbook*, Methuen, London.

Rhind D., Mounsey H. and Shepherd J. (1984) Automated mapping, *Proceedings of the Census of Population and Associated Social Statistics Conference*, 22 November, Sheffield.

Riggleman J.R. (1936) *Graphical Methods for Presenting Business Statistics*, McGraw-Hill, New York.

Rucker R. (1984) *The Fourth Dimension: Towards a Geometry of Higher Reality*, Houghton Mifflin Co., Boston, Massachusetts.

Salt J. (1990) Organisational labour migration: theory and practise in the United Kingdom, in *Labour Migration: The Internal Geographical Mobility of Labour in the Developed World* (eds J.H. Johnson and J. Salt), David Fulton Publishers, London.

Savage M. (1989) Spatial differences in modern Britain, in *Restructuring Britain: The Changing Social Structure* (eds C. Hamnett, L. McDowe and P. Sarre), Sage Publications and The Open University, London.

Scott J. (1989) The corporation and the class structure, in *Divided Nation: Social and Cultural Change in Britain* (eds L. McDowell, P. Sarre and C. Hamnett), Hodder and Stoughton and The Open University, London.

SEEDS (1987) *The South–South Divide*, The SEEDS Association, Harlow, Britain.

Selvin S., Merrill D., Sacks S., Wong L., Bedell L. and Schulman J. (1984) Transformations of maps to investigate clusters of disease, Lawrence Berkeley Lab Report LBL-18550.

Shepherd J., Westaway J. and Lee T. (1974) *A Social Atlas of London*, Clarendon Press, Oxford.

Sibert J.L. (1980) Continuous-colour choropleth maps, *Geo-Processing*, **1**, 207–217.

Skoda L. and Robertson J.C. (1972) Isodemographic map of Canada, Geographical Paper No. 50, Department of the Environment, Ottawa, Canada.

Smith D.R. and Paradis A.R. (1989) Three-dimensional GIS for the earth sciences, in *GIS: Three Dimensional Applications in Geographic Information Systems* (ed. J. Raper), Taylor & Francis, London.

Staudhammer J. (1975) Multi-dimensional function display using color scale, *Siggraph*, **9** (1), 181–183.

Szegö J. (1984) *A Census Atlas of Sweden*, Statistics Sweden, Central Board of Real Estate Data, Swedish Council for Building Research and The University of Lund, Stockholm.

Szegö J. (1987) *Human Cartography: Mapping the World of Man*, Swedish Council for Building Research, Stockholm.

Taylor P.J. (1978) Political geography, *Progress in Human Geography*, 2, 153–162.

Taylor D.R.F. (1985) The educational challenges of a new cartography, *Cartographica*, **22** (4), 19–37.

Taylor P.J. (1991) A future for geography, *Terra*, **103** (1), 21–31.

Thompson J.M. (1988) Advanced scientific visualization productivity with integrated analysis, graphics, and image processing techniques, *Electronic Imaging '88, International Electronic Imaging Exposition and Conference*, Boston, Massachusetts, pp. 1084–1085.

Thornthwaite C.W. and Slentz H.I. (1934) *Internal Migration in the United States*, University of Pennsylvania Press, Philadelphia.

Tobler W.R. (1959) Automation and cartography, *The Geographical Review*, **49**, 526–536.

Tobler W.R. (1961) Map transformations of geographic space, PhD, Department of Geography, University of Washington.

Tobler W.R. (1968) Transformations, in The Philosophy of Maps (ed. J.D. Nystuen), Discussion Paper No. 12, Michigan Inter-University Community of Mathematical Geographers, Wayne State University, pp. 2–4.

Tobler W.R. (1969) Geographical filters and their inverses, *Geographical Analysis*, **1**, 234–251.

Tobler W.R. (1973a) A continuous transformation useful for districting, *Annals of the New York Academy of Sciences*, **219**, 215–220.

Tobler W.R. (1973b) Choropleth maps without class intervals?, *Geographical Analysis*, **3**, 262–265.

Tobler W.R. (1986) Pseudo-cartograms, *The American Cartographer*, **13** (1), 43–50.

Tobler W.R. (1987) Experiments in migration mapping by computer, *The American Cartographer*, **14** (2), 155–163.

Tobler W.R. (1989a) Frame independent spatial analysis, in *The Accuracy of Spatial Databases* (ed. M. Goodchild and S. Gopal), Taylor & Francis, Philadelphia.

Tobler W.R. (1989b) An update to 'numerical map generalization', in Numerical generalization in cartography (ed. R.B. McMaster), *Cartographica*, **26** (1), 7–25 (University of Toronto Press).

Torguson J.S. (1990) Cartogram: a microcomputer program for the interactive construction of contiguous value-by-area cartograms, MA, University of Georgia.

Townsend P. with Corrigan P. and Kowarzik U. (1987) Poverty and labour in London: Interim Report of a centenary survey, Survey of Londoners' Living Standards No. 1, Low Pay Unit and Poverty Research Trust, London.

Tufte E.R. (1983) *The Visual Display of Quantitative Information*, Graphics Press, Cheshire, Connecticut.

Tufte E.R. (1988) Comment on: dynamic graphics for data analysis (1987); Becker, Cleveland and Wilks, in *Dynamic Graphics for Statistics* (eds W.S. Cleveland and M.E. McGill), Wandsworth, California.

Tufte E.R. (1990) *Envisioning Information*, Graphics Press, Cheshire, Connecticut.

Tukey P.A. and Tukey J.W. (1981) Graphical display of data sets in 3 or more dimensions, in *Interpreting Multivariate Data* (ed. V. Barnett), Part III, pp. 189–275.

Upton G.J.G. (1976) Research notes: diagrammatic representation of three-party contests, in *Political Studies*, vol. XXIV, Clarendon Press, pp. 448–454.

van Dreil J.N. (1989) Three-dimensional display of geologic data, in *GIS: Three Dimensional Applications in Geographic Information Systems* (ed. J. Raper), Taylor & Francis, London.

Walker F.A. (1870) *Statistical Atlas of the United States, Based on the Results of the Ninth Census, 1870*, US Bureau of the Census, Washington D.C.

Warnes A.M. and Law C.M. (1984) The elderly population of Great Britain: locational trends and policy implications, *Transactions of the Institute of British Geographers*, **9**, 37–59.

Warntz W.W. (1973) First variations on the theme, cartographics is to geographical science as graphics is to science generally, *Applied Geography and the Human Environment*, **2**, 54–85.

Warntz W.W. (1975) The pattern of patterns: current problems of sources of future solutions, in *Human Geography in a Shrinking World* (eds R. Abler, D. Janelle, A. Philbrick and J. Sommer), Duxbury Press, Massachusetts.

White R.D. (1984) The structural complexity of multivariate maps in national atlases, MA, Boston University.

Wilkie R.W. (1976) Maps and cartograms, *Statistical Abstracts of Latin America*, **17**, 1–23.

Williams R.L. (1976) The misuse of area in mapping census-type numbers, *Historical Methods Newsletter*, **9** (4), 213–216.

Worth C. (1978) Determining a vertical scale for graphical representations of three-dimensional surfaces, *The Cartographic Journal*, **15** (2), 86–92.

Wright P. (1989) The ghosting of the inner city, in *Divided Nation: Social and Cultural Change in Britain* (eds L. McDowell, P. Sarre and C. Hamnett), Hodder and Stoughton and The Open University, London.

Author Index

Subject Index

The Visualization of Spatial Social Structure, First Edition. Daniel Dorling.
© 2012 John Wiley & Sons, Ltd. Published 2012 by John Wiley & Sons, Ltd.

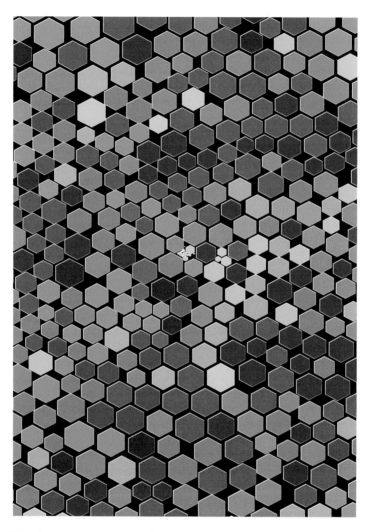

Inset of local voting patterns within London, 8× magnification of Figure 4.21. The full image and key are shown on page 129. The very small hexagons are the wards of the ancient City of London for which democratic local elections were and still are not held. Even on an equal population cartogram these very small areas are extremely hard to see. All hexagons are drawn with size in proportion to population and coloured to show local election results, 1987–1990.

Inset of national voting, housing and employment patterns within and around London, 4× magnification of Figure 8.10. The full image and key are shown on page 246. Each parliamentary constituency is represented by a face drawn with its size in proportion to population, cheeks in proportion to housing prices, smiling if unemployment was low, with large eyes if industry was 'young', and a wide nose if voting turnout was high, coloured to show general election results in 1987.